STUDENT
SOLUTIONS MANUAL

for

Descriptive
Inorganic Chemistry
Third Edition

Geoff Rayner-Canham
Tina Overton

W. H. Freeman and Company
New York

ISBN 0-7167-3759-0 (EAN: 9780716737599)

Printed in the United States of America

Second printing

W. H. Freeman and Company
41 Madison Avenue
New York, NY 10010

www.whfreeman.com

Table of Contents

Chapter 1

THE ELECTRONIC STRUCTURE OF THE ATOM: A REVIEW

Exercises

1.1   (a) Surface where the electron probability is zero.

(b) No two electrons in an atom can have exactly the same set of quantum numbers.

(c) Paramagnetic—the attraction into a magnetic field by an unpaired electron.

1.3

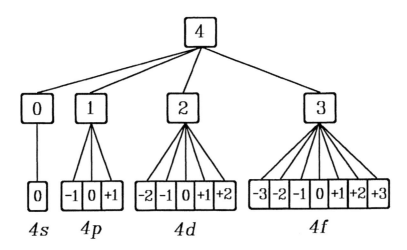

1.5 $5p$.

1.7   The quantum number n relates to the size of an orbital.

1.9   The two electrons paired and occupying the same orbital would be least favorable, because the pairing energy would be necessary to overcome the repulsive forces. Paired but in different orbitals also comes with an energy cost, as there is a finite probability that the electrons will occupy the same volume of space, again resulting in a repulsive energy factor. With parallel spins there is a zero probability of occupying the same volume of space; hence this is the lowest energy condition.

1.11   (a) [Ne]$3s^1$; (b) [Ar]$4s^23d^8$; (c) [Ar]$4s^13d^{10}$.

1.13 (a) [Ar]; (b) [Ar]; (c) $[Ar]3d^9$.

1.15 1+ and 3+. Thallium has a noble gas core ground-state electron configuration of $[Xe]6s^2 4f^{14} 5d^{10} 6p^1$. The $6p$ electron is lost first, giving an ion of 1+ charge; the two $6s$ electrons are lost next, giving an ion of 3+ charge.

1.17 1+. Silver has a noble gas core ground state electron configuration of $[Kr]5s^1 4d^{10}$. The $5s$ electron is lost first, giving an ion of 1+ charge.

1.19 (a) 2; (b) 0; (c) 4.

2p ⇅ ↑ ↑   2s ⇅

(a)   (b)

4s ↑   3d ↑ ↑ ↑ ↑ ↑

(c)

1.21 Electron configuration of atom: $[Rn]7s^2 5f^{14} 6d^{10} 7p^1$.
Electron configuration of +1 ion: $[Rn]7s^2 5f^{14} 6d^{10}$.
Electron configuration of +3 ion: $[Rn]5f^{14} 6d^{10} 7p$.

Beyond the Basics

1.23 9, 5, 121.

1.25 There are seven f orbitals. There are (at least) two separate ways of depicting them and designating them: the general set, and the cubic set. The seven solutions for the cubic set are: x^3, y^3, z^3, xyz, $z(x^2 - y^2)$, $y(z^2 - x^2)$, and $x(z^2 - y^2)$. The f_{x3}, f_{y3}, f_{z3} resemble the d_{z2} in that they have lobes along the particular axis, but with double "doughnut" rings around the middle rather than single rings. The other f orbitals resemble the d_{x2-y2}, d_{xy}, d_{xz}, and d_{yz} orbitals in that they consist of eight lobes (rather than four) between the axes. Several texts, such as that by Huheey, discuss f orbitals. See also E.A. Ogryzlo, On the Shapes of f Orbitals, *J. Chem. Educ.*, **42**, 150–151 (1965) and refs. therein.

1.27 Of course, one can argue that orbitals are human constructs only! This question is a good topic for debate, but these authors veers toward the view that an orbital actually exists only when it is populated. Empty orbitals only potentially exist.

Chapter 2

An Overview of the Periodic Table

Exercises

2.1   (a) Elements scandium, yttrium, and lanthanum through lutetium.

(b) Apparent radius of an atom when it is in non-bonded contact with another atom;

(c) Actual nuclear charge experienced by a electron in an atom.

2.3   There was no space for argon because it did not fit in one of the then-known groups (the noble gas group was then unknown). Also, because Mendeleev's table was based on atomic mass, argon (39.9) should have been placed between potassium (39.1) and calcium (40.1).

2.5   The long form of the table correctly depicts the order of elements; the disadvantage is that the table becomes very elongated.

2.7   The *-ium* ending is used normally to indicate a metal. The ending *-on* has been used for non-metals. When helium was discovered by spectroscopy, it was assumed to be a metal, and hence the wrong suffix was chosen and never changed.

2.9   With nuclei up to 26 protons, nuclear fusion is an exothermic process and is thus favored. Beyond this point, fusion is endothermic and requires the energy from a supernova explosion to happen.

2.11   (a) Bismuth; (b) technetium; (c) bromine.

2.13   Sodium, as it has an odd number of protons (according to isotope tables, sodium has one stable isotope, that with 12 neutrons—an even number—while magnesium has three stable isotopes).

2.15   50. The number of neutrons must be greater than the number of protons. Fifty is the next higher "magic number" above 39, the proton number.

2.17   (a) Several non-metals, such as iodine, and some compounds, such as FeS_2, fool's gold, have metallic lusters.

(b) Diamond, a carbon (nonmetal) allotrope, has the highest thermal conductivity of all substances.

(c) The other common carbon allotrope, graphite, also a nonmetal, is a good electrical conductor in two dimensions.

2.19   Potassium. As the effective atomic charge on the outermost electrons increases across a period, so the covalent radius will decrease, resulting in a smaller radius for calcium.

2.21   With the poorly shielding $3d$ orbitals having been filled corresponding to the added protons, the effective nuclear charge on the outer, $4p$, electrons will be increased; hence the covalent radius will decrease.

2.23   Table of effective nuclear charge for the second period elements using Slater's rules.

Element	Li	Be	B	C	N	O	F	Ne
Z	3	4	5	6	7	8	9	10
$1s$	2.65	3.65	4.65	5.65	6.65	7.65	8.65	9.65
$2s$	1.30	1.95	2.60	3.25	3.90	4.55	5.20	5.85
$2p$			2.60	3.25	3.90	4.55	5.20	5.85

The differences between the simplistic Slater's rules and the more sophisticated Clementi and Raimondi method are quite small. The disadvantage of the Slater method is that it does not distinguish between s and p electrons in terms of shielding.

2.25   Effective nuclear charge on $4s = 2.95$; $3d = 5.60$.

2.27   Phosphorus. With increasing nuclear charge across the period, the ionization energy will increase.

2.29 Group 2. The size of the ionization energies increases rapidly between the second and third values. Hence the third and subsequent values must correspond to the ionization of inner electrons. Two outer electrons indicate the element belongs to the alkaline earth metals (Group 2).

2.31 Electron configurations: Na = $1s^2 2s^2 2p^6 3s^1$; Mg = $1s^2 2s^2 2p^6 3s^2$. As the $2s$ electrons of magnesium experience a higher effective nuclear charge than that of sodium, magnesium will have the higher first ionization energy. For sodium, the second electron to be removed will come from the inner $2p$ orbitals. Hence it will be sodium that has the higher second ionization energy. In both cases, the third electron to be removed will come from the $2p$ orbitals. The $2p$ electrons of magnesium will experience the higher effective nuclear charge. Thus magnesium will have the higher third ionization energy.

2.33 Positive. With an electron configuration of $1s^2$, any gained electron would add to the $2s$ orbital. This would experience stronger repulsions from the inner $1s$ electrons than attraction from the well-shielded nucleus. Hence it would be an endothermic process.

2.35 In each case, the preferred isotope will be that having a full shell of neutrons (126). Hence the answers are (a) 208; (b) 209; (c) 210.

Beyond the Basics

2.37 The average kinetic energy (= ½mv^2) of all gases are the same under identical conditions of temperature. As dihydrogen and helium have such low molecular masses, their average velocities will be much higher. A sufficient number of molecules will have greater energy than the escape velocity for them to escape from the Earth's gravitational field. Over time, the vast majority of these molecules have escaped.

2.39 Element 117—let us call it X—would be a member of the halogens. It should be a solid at room temperature and exist as X_2 molecular units. It should form an anion X⁻, but at the same time, X is on the border with metals, so positive oxidation states, particularly +1 and +3, should be quite common.

Chapter 3

COVALENT BONDING

Exercises

3.1 (a) Acronym for the **l**inear **c**ombination of **a**tomic **o**rbitals and it is a sophisticated theory for the representation of bonding in covalent molecules.

(b) A molecular orbital in which the increased electron density lies between the two nuclei involved in the bonding.

(c) The acronym for valence shell electron pair repulsion, a method of predicting molecular shape using simple electron pair concepts of bonding.

(d) Hybridization involves the concept of mixing atomic orbitals on a central atom to give hybrid orbitals that have the directions to give the maximum overlap with the atomic orbitals of the surrounding atoms.

3.3 The bond order would be ½ and the ion would be paramagnetic.

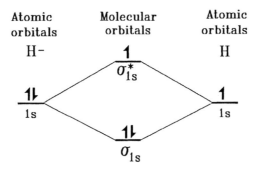

3.5   Bond order 2½. Electron configuration: $(\sigma_{2s})^2(\sigma*_{2s})^2(\pi_{2p})^4(\pi_{2p})^1$.

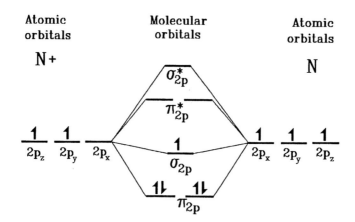

3.7   Triple bond.

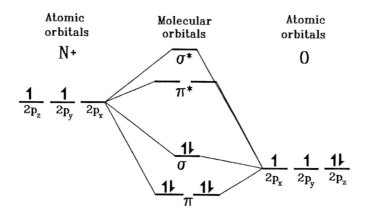

3.9   Single bond. The ordering for the earlier elements predicts two unpaired electrons, while that for the later elements predicts paired electrons. Hence if the former ordering is correct, the molecule should be paramagnetic, but the latter ordering would be diamagnetic.

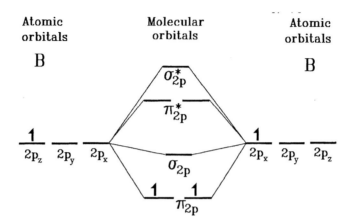

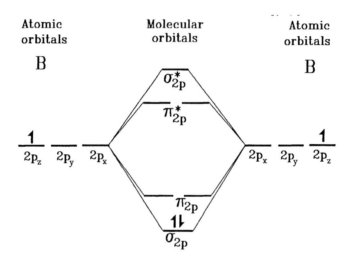

3.11   Electron dot diagrams:

(a)   (b)

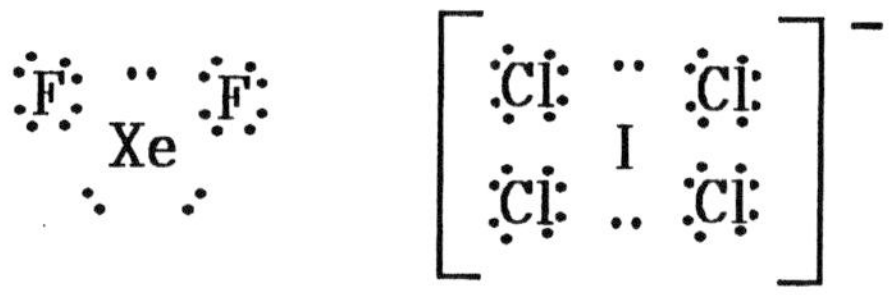

(c)   (d)

3.13   Electron dot diagram (a), resonance structures (b), and partial bond representation (c).

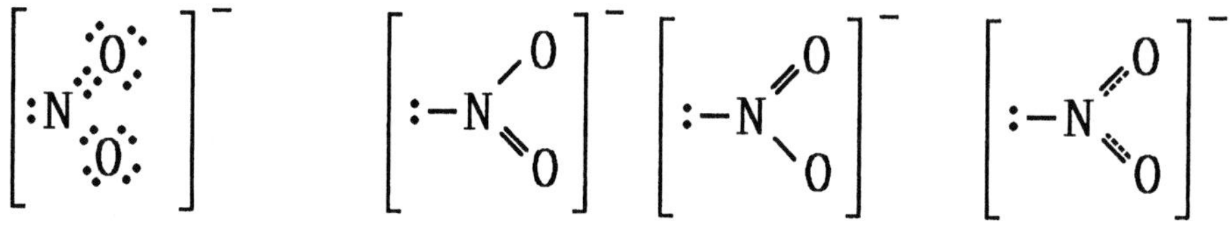

(a)   (b)   (c)

3.15   Electron dot diagrams (a), formal charge structures (b), partial bond representation (c).

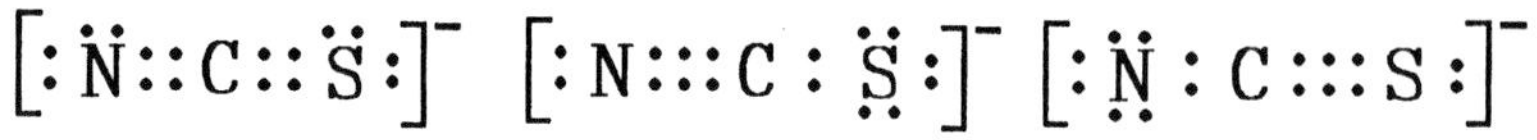

(a)

formal
charge

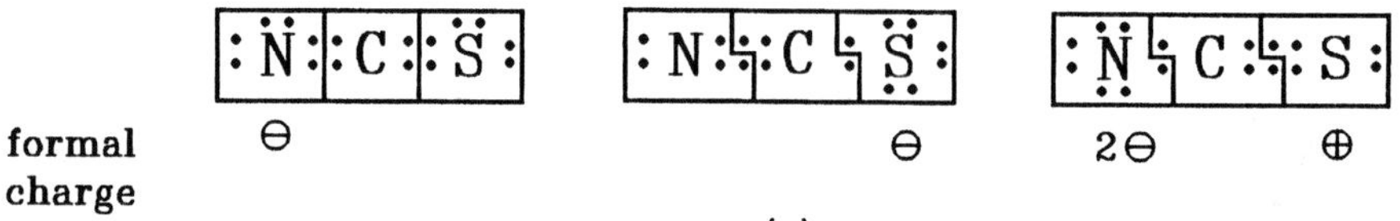

(b)

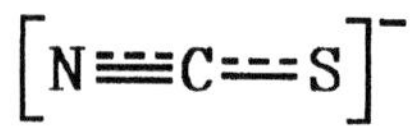

(c)

3.17   (a) Tetrahedral, vee-shaped; (b) Tetrahedral, trigonal pyramidal;
        (c) Trigonal bipyramidal, linear; (d) Octahedral, square planar.

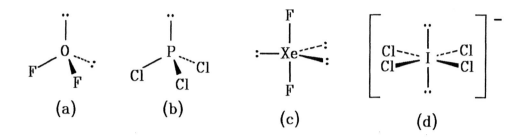

3.19   Linear: carbon disulfide and xenon difluoride; vee-shaped: chlorine dioxide
        (<109½°) [in fact, the bonding is not so simple, as we discuss in Chapter 16],
        tin(II) chloride (<120°), and nitrosyl chloride (<120°).

3.21   Oxygen difluoride and phosphorus trichloride. The bond angles will be reduced
        as a result of the presence of the lone pairs.

3.23   (a) sp^3; (b) sp^3; (c) sp^3d; (d) sp^3d^2.

3.25 The hybrid orbitals used would be *sp*, accounting for the linear arrangement.

2p — — —	2p <u>1</u> — —	2p — —	2p — —
		sp <u>1</u> <u>1</u>	sp <u>1↓</u> <u>1↓</u>
2s <u>1↓</u>	2s <u>1</u>		
(a)	(b)	(c)	(d)

3.27 Hydrogen selenide, for with more electrons, the dispersion forces would be greater (the slightly weaker dipole-dipole interaction than that of hydrogen sulfide is a less important factor).

3.29 Polar: oxygen difluoride and phosphorus trichloride; non-polar: xenon difluoride and the tetrachloroiodate ion.

3.31 Ammonia, as neighboring molecules will hydrogen-bond to each other, providing a strong intermolecular force and hence a higher boiling point than phosphine, which has the weaker dipole-dipole interactions.

3.33 The possible hybridization: (a) *sp*; (b) sp^2; (c) sp^3; (d) sp^3d; (e) sp^3d^2.

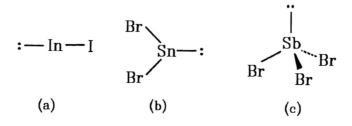

        (a)   (b)   (c)

$$:—\overset{\displaystyle Cl}{\underset{\displaystyle Cl}{Te}}\overset{---Cl}{\underset{Cl}{}}$$

(d)

$$\overset{\displaystyle F}{\underset{\displaystyle ..}{F-\!-\!-I-\!-\!-F}}$$

(e)

Beyond the Basics

3.35   The following figures show the overlap of an *s* orbital with a typical *d* orbital to
form a σ bond; overlap of a *p* orbital with a typical *d* orbital to form a σ bond; and
overlap of a *p* orbital with a typical *d* orbital to form a π bond. Black dots
indicate the location of the nuclei.

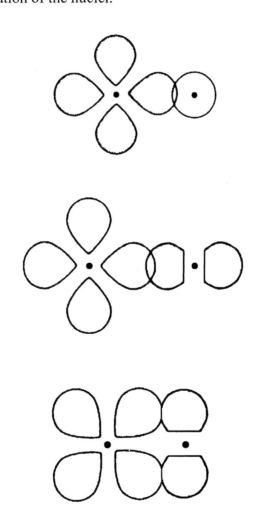

3.37 Using formal charge structures, we see that the NNO arrangement provides two possibilities with only one formal charge per atom. The two feasible formal charge structures for NON both have double charges.

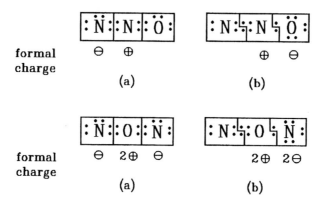

formal
charge

(a)   (b)

formal
charge

(a)   (b)

3.39 The two feasible formal charge electron arrangements for the CON⁻ ion have five formal charges! This is a very unlikely structure.

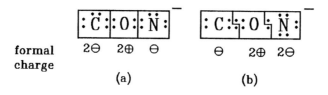

formal
charge

(a)   (b)

3.41 Let us look at the bond order of the product molecules in each case. NO⁺ and CN⁻ are both triply bonded while NO⁻ and CN⁺ will be double-bonded. The former combination will therefore be energetically preferred.

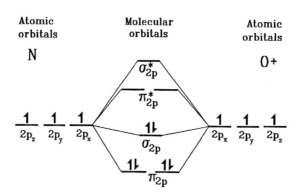

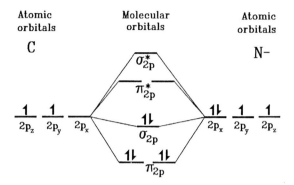

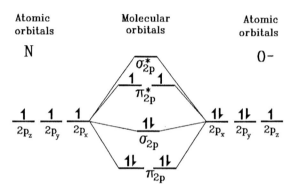

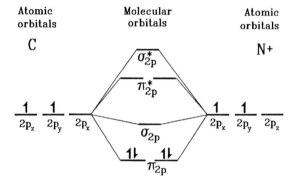

3.43   One would expect the halide with the larger number of halogen atoms to have the higher boiling point as a result of the greater number of electrons and hence the stronger dispersion forces. A possible answer (but not necessarily the correct one!) would be that antimony has an electronegativity comparable to some metals. As a result, there may be a significant ionic character to the bonding in $SbCl_3$ (the lower oxidation state and hence lower charge density species) resulting in a higher boiling point than otherwise expected.

Chapter 4

METALLIC BONDING

Exercises

4.1   (a) Model in which it is assumed the metal consists of metal ions with free electrons.

(b) Smallest fragment of a crystal lattice that, if repeated, will recreate the entire structure.

(c) Combination of two or more solid metals.

4.3   High electrical conductivity, high thermal conductivity, high reflectivity, and high boiling point (any three).

4.5   The overlap of the $3s$ and $3p$ bands means that electrons in the full $3s$ band can "spill over" into the $3p$ band enabling electron mobility through the crystal lattice.

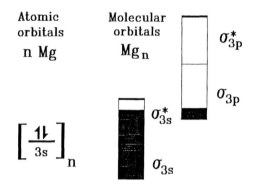

4.7   For metallic behavior, the orbitals of the atoms must overlap. In the gas phase, metal atoms are moving freely as single atoms or as discrete molecules, such as dilithium, Li_2.

4.9   Cubic and hexagonal, hexagonal being the closer packed.

4.11   The simple cubic unit cell contains $4 \times \frac{1}{4}$ atoms; that is, one atom.

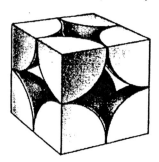

4.13   The atoms must be about the same size, they must adopt the same structure, and they must have similar chemical properties.

Beyond the Basics

4.15   If we take the radius of an atom as r, then the side length of a unit cell for the simple cubic lattice will be $2r$. The volume of the atom will be $\frac{4}{3}\pi r^3$, while the volume of the cube will be $(2r)^3$. The ratio of these gives 0.52. Thus the empty space, expressed as a percentage will be 48%.

4.17   The face diagonal length will be $4r$. Using Pythagoras' Theorem, the length of the unit cell edge will be $[4/(2)^{\frac{1}{2}}]r = 2.83r$.

4.19   (a) Using the answer to 4.18, the metallic radius of chromium will be 288/2.31 pm = 125 pm.

(b) The density can be calculated from $\dfrac{\text{unit cell mass}}{\text{unit cell volume}}$.

The unit cell volume will be $(288 \times 10^{-10} \text{ cm})^3 = 2.39 \times 10^{-23} \text{ cm}^3$

Each unit cell contains two atoms, thus mass $= \dfrac{2 \times 52.0 \text{ g} \cdot \text{mol}^{-1}}{6.02 \times 10^{23} \text{ mol}^{-1}} = 1.73 \times 10^{-22}$ g.

Density $= \dfrac{1.73 \times 10^{-22} g}{2.39 \times 10^{-23} cm^3} = 7.24$ g·cm^{-3}.

4.21   A face-centered cubic unit cell contains four atoms.

Thus mass $= \dfrac{4 \times 107.9 \text{ g} \cdot \text{mol}^{-1}}{6.02 \times 10^{23} \text{ mol}^{-1}} = 7.17 \times 10^{-22}$ g

Volume $= \dfrac{7.17 \times 10^{-22} \text{ g}}{10.50 \text{ g} \cdot \text{cm}^{-3}} = 6.83 \times 10^{-23}$ cm^3 $= 6.83 \times 10^7$ pm^3

Length of side $= \sqrt[3]{(6.83 \times 10^7 \text{ pm}^3)} = 409$ pm

Using result from 4.17, radius of silver atom $= (409 \text{ pm})/(2.83) = 145$ pm

4.23   The crucial questions: Is there a hazard? What is the evidence? What are the alternatives? Are the alternatives as long-lasting? Is it safer to leave in old mercury amalgam fillings rather than remove them?

Chapter 5

Ionic Bonding

Exercises

5.1 (a) Distortion of an anion from a spherical shape.

(b) Holes between anions in the crystal packing in which cations can fit.

(c) Diagram used to show that many elements and compounds possess combinations of the three bonding categories: metallic, covalent, and ionic.

5.3 Hard and brittle crystals; high melting points; electrically conducting in liquid phase and in aqueous solution.

5.5 (a) K^+ as the radius will be determined by the inner orbitals ($2s$ and $2p$) while the radius of K is determined by the $3s$ orbital.

(b) Ca^{2+} as the ions are isoelectronic but calcium has one more proton, hence a higher Z_{eff} and smaller ionic radius.

(c) Rb^+ as again the ions are isoelectronic with rubidium having two more protons than bromide, hence a higher Z_{eff} and smaller ionic radius.

5.7 NaCl, as chloride is smaller than iodide; hence the charge is more concentrated (charge density is higher) and the ionic attraction will be stronger in NaCl than in NaI. The stronger the ionic attraction, the higher the temperature needed to melt the ionic lattice.

5.9 Ag^+, as it has the lowest charge density.

5.11 Tin(II) chloride has a higher melting point as tin(II) has a fairly low charge density. The bonding in $SnCl_2$ will be predominantly ionic. The high charge density of the tin(IV) ion would result in covalent bonding for $SnCl_4$ and therefore a low melting point.

5.13 No, ionic compounds do not dissolve in non-polar solvents because the dissolving process requires the formation of ion-dipole interactions. Molecules of a non-polar solvent do not possess dipoles.

5.15   Magnesium chloride, as the dipositive smaller magnesium ion has a significantly higher charge density than the monopositive larger sodium ion.

5.17   Ions are assumed to be charged, incompressible, nonpolarizable spheres; ions try to surround themselves with as many ions of the opposite charge as possible; the overall cation-anion ratio must reflect the stoichiometry of the compound.

5.19   In general, the anions are much larger than the cations; hence it is more appropriate to consider the lattice as an array of anions with the smaller cations fitting the interstices.

5.21

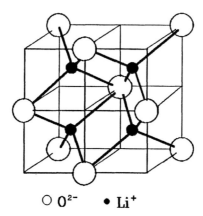

○ O^{2-} ● Li^+

5.23   (a) Metallic and a lesser contribution of ionic; (b) covalent and a lesser contribution of ionic.

5.25   The choice would be metallic or covalent. With such a low melting point (and without seeing the element), the first choice would be covalent. A measurement of its high electrical conductivity (and observation of the metallic luster) would confirm it to have metallic bonding.

Beyond the Basics

5.27 IF_7—the compound does exist. It is probable that the chlorine atom is too small for seven iodine atoms to fit around it.

5.29   In the solid, there is shared ionic bonding with the six nearest neighbors of each ion. In the gas phase, there is direct ionic/covalent bonding between pairs of ions.

5.31   (a) Copper(II) chloride. As both copper ions are in low oxidation states, covalency will not be a major factor. The higher charge density copper(II) ion will result in a higher lattice energy in its chloride and therefore a higher melting point.
(b) Lead(II) chloride. The very high charge density of the lead(IV) ion will result in the compound having essentially covalent bonding and therefore a low melting point. Ionic lead(II) chloride will have a high melting point.

5.33   The diagonal length through the center of the unit cell will be $2(r_+ + r_-)$. Using Pythagoras' Theorem, the length of the unit cell edge will be $[2/(3)^{1/2}](r_+ + r_-) = 1.15(r_+ + r_-)$.

5.35   The unit cell of rubidium chloride contains four pairs of ions.

Thus mass $= \dfrac{4 \times 121.0 \text{ g} \cdot \text{mol}^{-1}}{6.02 \times 10^{23} \text{ mol}^{-1}} = 8.04 \times 10^{-22}$ g

Volume $= \dfrac{8.04 \times 10^{-22} \text{ g}}{2.76 \text{ g} \cdot \text{cm}^{-3}} = 2.91 \times 10^{-22}$ cm^3 $= 2.91 \times 10^8$ pm^3

Length of side $= \sqrt[3]{(2.91 \times 10^8 \text{ pm}^3)} = 663$ pm

Sum of ionic radii $= (663 \text{ pm})/2 = 331$ pm

Thus $r(\text{Rb}^+) = 331$ pm $- 167$ pm $= 164$ pm

5.37   The unit cell of sodium hexafluoroantimonate(V) contains four pairs of ions.

Thus mass $= \dfrac{4 \times 258.8 \text{ g} \cdot \text{mol}^{-1}}{6.02 \times 10^{23} \text{ mol}^{-1}} = 1.72 \times 10^{-21}$ g

Volume $= \dfrac{1.72 \times 10^{-21} \text{ g}}{4.37 \text{ g} \cdot \text{cm}^{-3}} = 3.94 \times 10^{-22}$ cm^3 $= 3.94 \times 10^8$ pm^3

Length of side $= \sqrt[3]{(3.94 \times 10^8 \text{ pm}^3)} = 733$ pm

Sum of ionic radii $= (733 \text{ pm})/2 = 367$ pm

Thus $r(\text{SbF}_6^-) = 367$ pm $- 116$ pm $= 251$ pm

5.39 Fe_2O_3, RuO_4, OsO_4; +3, +8, +8; the highest oxidation number is the same as the Group number.

Chapter 6

INORGANIC THERMODYNAMICS

Exercises

6.1   (a) Reaction that occurs without external "help," or a reaction for which ΔG° is negative.

(b) Measure of disorder.

(c) Enthalpy change when a mole of a substance is formed from its constituent elements in their standard phases at 298 K and 100 kPa.

6.3   The entropy change is probably negative as according to the chemical equation, there is a decrease by one half mole of gas (gases have a much higher entropy than liquids and solids).

$$Ca(s) + \tfrac{1}{2}\,O_2(g) \rightarrow CaO(s)$$

If the entropy change is negative, then for a spontaneous reaction, the enthalpy change must be negative.

6.5 $\Delta H_f^\circ = -286 \text{ kJ·mol}^{-1}$;

$\Delta S_f^\circ = [1 \times (70) - 1 \times (131) - \tfrac{1}{2} \times (205)]$ J·mol^{-1}·K^{-1}

$= -163$ J·mol^{-1}·K^{-1} $= -0.163$ kJ·mol^{-1}·K^{-1}

$\Delta G_f^\circ = -286$ kJ·mol^{-1} $- (298\text{K})(-0.163$ kJ·mol^{-1}·K$^{-1})$

$= -237$ kJ·mol^{-1}

The reaction is spontaneous at SATP.

6.7 $PCl_5(g) + SO_3(g) \rightarrow SO_2Cl_2(g) + POCl_3(l)$

$\Delta G^\circ = \Sigma \Delta G_f^\circ(\text{products}) - \Sigma \Delta G_f^\circ(\text{reactants})$

$\Delta G^\circ = [1 \times (-314) + 1 \times (-521)] - [1 \times (-305) + 1 \times (-371)]$ kJ·mol^{-1}

$= -159$ kJ·mol^{-1}

This is only an approximate value of ΔG° as the ΔG_f° values are not those of the common phases at SATP.

6.9   The N=N bond will be stronger as multiple bonds are stronger than single bonds between the same elements.

6.11 Bonds broken = 4(S – S) + 8(H–S) = [(4 × 266) + (8 × 363)] kJ·mol^{-1}
      = 3968 kJ·mol^{-1}
      Bonds formed = 8(S – S) + 4(H–H) = [(8 × 266) + (4 × 432)] kJ·mol^{-1}
      = 3856 kJ·mol^{-1}
      Approximate enthalpy of reaction = [3968 – 3856] kJ·mol^{-1} = +112 kJ·mol^{-1}

6.13 Sodium chloride, lithium fluoride, magnesium oxide. LiF will be higher than NaCl as the former has smaller, higher charge density ions. MgO will be the highest as the charges (and thus charge densities) are higher.

6.15 The first two terms will be [8/(3)^{½} – 6/2] = 1.62. Again, more terms must be considered to approach the limiting value of 1.763.

6.17 $$U = -\frac{(6.02 \times 10^{23} \text{ mol}^{-1}) \times 2.519 \times 2 \times 1 \times (1.602 \times 10^{-19} \text{ C})^2}{4 \times 3.142 \times (8.854 \times 10^{-12} \text{ C}^2 \cdot \text{J}^{-1} \cdot \text{m}^{-1})(2.31 \times 10^{-10} \text{ m})}\left(1 - \frac{1}{8}\right)$$
      = –2649 kJ·mol^{-1}

6.19

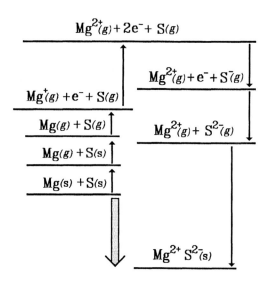

6.21 Na(s) sublimation = +107 kJ·mol^{-1}

½ H$_2$ bond energy = +216 kJ·mol^{-1}

Na(g) 1st ionization energy = +502 kJ·mol^{-1}

H(g) electron affinity (EA H) = to find

Na$^+$H$^-$ lattice energy = −782 kJ·mol^{-1}

ΔH_f^o(NaCl(s)) = −411 kJ·mol^{-1}

−411 kJ·mol^{-1} = [+107 + 216 +502 + (EA H) − 782] kJ·mol^{-1}

(EA H) = −454 kJ·mol^{-1}

6.23 Because of the high charge density of the oxide, O^{2-}, ion, the lattice energy of oxides will be very high. This will more than compensate for the fact that the net electron affinity is positive. It is the importance of the lattice energy factor that results in oxides only being stable in the solid phase.

Beyond the Basics

6.25 The term "the permittivity of free space" is a constant that relates the attractive force between two point charges. The mathematical formula is known as Coulomb's law and it has the formula

$$F = \frac{1}{4\pi\varepsilon_0}\frac{q_1 q_2}{r^2}$$

where F is the force, q_1 and q_2 are the magnitudes of the point charges, r is the distance between the charges, and ε_0 is the permittivity of free space. The permittivity of free space is an important constant throughout equations related to electric charge and electromagnetic waves. The most important relationship, perhaps, is:

$$c^2 = \frac{1}{\varepsilon_0 \mu_0}$$

where c is the velocity of light and μ_0 is the permeability of free space. Thus the velocity of light is related to these two other physical constants.

6.27 As can be seen, the enthalpy of formation for both $NaCl_2$ and $NaCl_3$ are positive. Even though the lattice energies resulting from the more highly charged ions are much greater, it is not enough to compensate for the ionization of one or two $2p$ (inner) electrons.

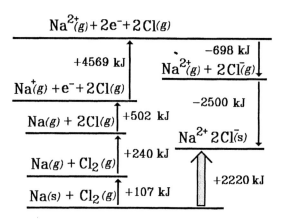

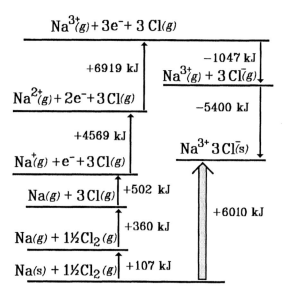

6.29 Magnesium oxide, being a 2+/2− ion combination will have a higher lattice energy than the 2+/1− ion combination of magnesium chloride. This energy input must be greater than the energy released from the hydration of the ions.

6.31   From the energy diagram, $\Delta H_f^\circ = -913 \text{ kJ·mol}^{-1}$, while the tabulated value is -933 kJ·mol^{-1}.

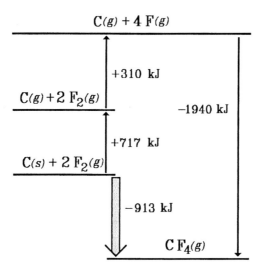

6.33   From the following energy diagram (using the ΔH_f° value from Appendix 2), the Cl–F bond energy is calculated to be 250 kJ·mol^{-1}.

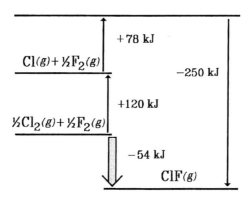

6.35   The lattice energy of the calcium chloride will also be much greater. This will negate the higher hydration energy.

6.37   For calcium sulfate:

$-17.8 \text{ kJ} \cdot \text{mol}^{-1} = 2653 \text{ kJ} \cdot \text{mol}^{-1} - (1650 + \Delta H_{hydr}(SO_4^{2-})) \text{ kJ} \cdot \text{mol}^{-1}$

$\Delta H_{hydr}(SO_4^{2-}) = -1021 \text{ kJ} \cdot \text{mol}^{-1}$

For strontium sulfate:

$-8.7 \text{ kJ} \cdot \text{mol}^{-1} = 2603 \text{ kJ} \cdot \text{mol}^{-1} - (1480 + \Delta H_{hydr}(SO_4^{2-})) \text{ kJ} \cdot \text{mol}^{-1}$

$\Delta H_{hydr}(SO_4^{2-}) = -1132 \text{ kJ} \cdot \text{mol}^{-1}$

For barium sulfate:

$+19.4 \text{ kJ} \cdot \text{mol}^{-1} = 2423 \text{ kJ} \cdot \text{mol}^{-1} - (1360 + \Delta H_{hydr}(SO_4^{2-})) \text{ kJ} \cdot \text{mol}^{-1}$

$\Delta H_{hydr}(SO_4^{2-}) = -1044 \text{ kJ} \cdot \text{mol}^{-1}$

The values are consistent, within the errors of the data.

6.39   The cycle is shown in the diagram.

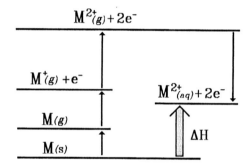

Using the values for magnesium:

$\Delta H = [+147 + 744 + 1457 - 1920] \text{ kJ} \cdot \text{mol}^{-1} = +428 \text{ kJ} \cdot \text{mol}^{-1}$

Using the values for lead:

$\Delta H = [+196 + 722 + 1457 - 1477] \text{ kJ} \cdot \text{mol}^{-1} = +898 \text{ kJ} \cdot \text{mol}^{-1}$

The only significant difference between the two ions is in their hydration enthalpies. If we look at the charge densities, we see that the charge density for the magnesium ion is much greater than that of the lead(II) ion. Hence there will be a stronger attraction (more exothermic reaction) between the magnesium ion and the oxygen ends of water molecules than for the equivalent process with the lead(II) ion (you can also argue using comparative ionic radii as the charges are the same).

6.41   Using the Kapustinskii equation:

$$U = -\frac{1.202 \times 10^5 \times 2 \times 1 \times 1}{348}\left(1 - \frac{34.5}{348}\right) = -622 \text{ kJ·mol}^{-1}$$

compared with -668 kJ·mol^{-1} experimentally and -636 kJ·mol^{-1} from the Born-Landé equation.

6.43   According to the Appendix, the NaCl-structure lattice energy of rubidium chloride
= 693 kJ·mol^{-1}. We can calculate the theoretical CsCl lattice energy using the Born-Landé equation:

$$U = -\frac{(6.02 \times 10^{23} \text{ mol}^{-1}) \times 1.763 \times 1 \times 1 \times (1.602 \times 10^{-19} \text{ C})^2}{4 \times 3.142 \times (8.854 \times 10^{-12} \text{ C}^2 \cdot \text{J}^{-1} \cdot \text{m}^{-1})(3.33 \times 10^{-10} \text{ m})}\left(1 - \frac{1}{9.5}\right)$$

$$= -657 \text{ kJ·mol}^{-1}$$

Thus approximate ΔH for the transformation = $[(-657) - (-693)]$ kJ·mol^{-1}
= $+36$ kJ·mol^{-1}

6.45   First, we have to calculate the lattice energy of NH_4F:

$$U = -\frac{(6.02 \times 10^{23} \text{ mol}^{-1}) \times 1.641 \times 1 \times 1 \times (1.602 \times 10^{-19} \text{ C})^2}{4 \times 3.142 \times (8.854 \times 10^{-12} \text{ C}^2 \cdot \text{J}^{-1} \cdot \text{m}^{-1})(2.56 \times 10^{-10} \text{ m})}\left(1 - \frac{1}{8}\right)$$

$$= -778 \text{ kJ·mol}^{-1}$$

Then we can construct a Born-Haber type cycle:

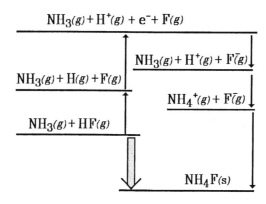

ΔH reaction can be calculated from the difference in the enthalpies of formation:

$\Delta H = [(-464) - (-273) - (-46)] \text{ kJ·mol}^{-1} = -145 \text{ kJ·mol}^{-1}$

The ionization energy of hydrogen ($+1537 \text{ kJ·mol}^{-1}$) is provided in the question.

The other information needed to complete the cycle, except for:

$NH_3(g) + H^+(g) + F^-(g) \rightarrow NH_4^+(g) + F^-(g)$

is found in the Appendices:

Bond energy (H–F) = $+565 \text{ kJ·mol}^{-1}$

Electron Affinity F = -328 kJ·mol^{-1}

$-145 \text{ kJ·mol}^{-1} =$

$\quad [(+565) + (+1537) + (-328) + (\text{proton affinity}) + (-778)] \text{ kJ·mol}^{-1}$

proton affinity = $-1141 \text{ kJ·mol}^{-1}$

6.47   Metals certainly do not "want to lose electrons," as is apparent from the high positive values of the ionization energies. It is generally but not always true that the electron affinities for nonmetals are negative, for example, the second electron affinity of oxygen is positive.

Chapter 7

ACIDS AND BASES

Exercises

7.1   (a) $H^+(aq) + OH^-(aq) \rightarrow H_2O(l)$

       (b) $2\ HCO_3^-(aq) + Co^{2+}(aq) \rightarrow CoCO_3(s) + H_2O(l) + CO_2(g)$

       (c) $OH^-(aq) + CH_3COOH(aq) \rightarrow CH_3COO^-(aq) + H_2O(l)$

7.3   (a) Pairs of species that differ in formula by one ionizable hydrogen.

       (b) Solvent that undergoes its own acid-base reaction, for example.

$$HA + HA \leftrightarrow H_2A^+ + A^-$$

       (c) Ability of a substance to act as an acid or a base, for example, water can act as a base to form the hydronium ion or as an acid to form the hydroxide ion.

7.5 $ClNH_2(aq) + H_2O(l) \leftrightarrow ClNH_3^+(aq) + OH^-(aq)$

7.5   (a) $NH_4^+(aq) + H_2O(l) \leftrightarrow NH_3(aq) + H_3O^+(aq)$

            (no reaction for the $NO_3^-(aq)$ ion)

       (b) $CN^-(aq) + H_2O(l) \leftrightarrow HCN(aq) + OH^-(aq)$

            (no reaction for the $K^+(aq)$ ion)

       (c) $HSO_4^-(aq) + H_2O(l) \leftrightarrow SO_4^{2-}(aq) + H_3O^+(aq)$

            (no reaction for the $Na^+(aq)$ ion)

7.7 $ClNH_2(aq) + H_2O(l) \leftrightarrow ClNH_3^+(aq) + OH^-(aq)$

7.9 $H_2SO_4(l) + H_2SO_4(l) \leftrightarrow H_3SO_4^+(H_2SO_4) + HSO_4^-(H_2SO_4)$

7.11   (a) The ammonium ion, NH_4^+; (b) the amide ion, NH_2^-.

7.13 $HF(H_2SO_4) + H_2SO_4(l) \leftrightarrow H_2F^+(H_2SO_4) + HSO_4^-(H_2SO_4)$

       HF is acting as a base and H_2F^+ is the conjugate acid,

       H_2SO_4 is acting as an acid and HSO_4^- is the conjugate base.

7.15 $HSeO_4^-$ (base), H_2SO_4 (conjugate acid); H_2O (acid), OH^- (conjugate base).

7.17 As the ionization process depends upon the breaking of the bond to hydrogen:
$$H_2X(aq) + H_2O(l) \leftrightarrow H_3O^+(aq) + HX^-(aq)$$
the weaker the H–X bond, the stronger the acid. Bond energies usually decrease down a group (see Bond Energy section in Chapter 6); hence the H–Se bond will be weaker than the H–S bond. Thus hydrogen selenide will be the stronger acid.

7.19 The hydrated ion will lose one hydrogen ion in a similar manner to the hydrated iron(III) and aluminum ions:
$$[Zn(OH_2)_6]^{2+}(aq) + H_2O(l) \leftrightarrow [Zn(OH_2)_5(OH)]^+(aq) + H_3O^+(aq)$$

7.21 As stepwise acid-base equilibria proceed to a lesser and lesser extent, the diprotic acid must be present in the least proportion.
$$H_2NNH_2(aq) + H_2O(l) \leftrightarrow H_2NNH_3^+(aq) + OH^-(aq)$$
$$H_2NNH_3^+(aq) + H_2O(l) \leftrightarrow {}^+H_3NNH_3^+(aq) + OH^-(aq)$$

7.23 (a) Acidic, as aluminum is a small high charge cation that will lose an ionizable hydrogen from one of the surrounding water molecules:
$$[Al(OH_2)_6]^{3+}(aq) + H_2O(l) \leftrightarrow [Al(OH_2)_5(OH)]^{2+}(aq) + H_3O^+(aq)$$
(b) Neutral, the sodium ion will stay unchanged, and the iodide ion is the conjugate base of a strong acid, so it, too, will remain unchanged.

7.25 With a smaller pK_b, A^- must be the stronger base. Hence HA must be the weaker acid. Thus HB will be the stronger acid.

7.27 $H_3PO_4(aq) + HPO_4^{2-}(aq) \leftrightarrow 2\ H_2PO_4^-(aq)$

7.29 (a) N_2O_5, (b) CrO_3, (c) I_2O_7.

7.31 (a) SiO_2 (acid), Na_2O (base); (b) NOF (acid), ClF_3 (base); (c) Al_2Cl_6 (acid), PF_3 (base).

7.33 (a) No effect

(b) Increasing pH (moderately large change)
$$Se^{2-}(aq) + H_2O(l) \leftrightarrow HSe^-(aq) + OH^-(aq)$$

(c) Decreasing pH (large change by analogy with aluminum ion)
$$[Sc(OH_2)_6]^{3+}(aq) + H_2O(l) \leftrightarrow [Sc(OH_2)_5(OH)]^{2+}(aq) + H_3O^+(aq)$$

(d) Increasing pH (small change)
$$F^-(aq) + H_2O(l) \leftrightarrow HF(aq) + OH^-(aq)$$

7.35 (a) Weakly basic; (b) neutral; (c) moderately basic; (d) strongly basic.

7.37 (a) Strongly basic; (b) very strongly basic

7.39 The chemical reaction is
$$MgO(s) + H_2O(l) \rightarrow Mg(OH)_2(s)$$
$$\Delta G° = [1(-834) - 1(-569) - 1(-237)] \text{ kJ·mol}^{-1}$$
$$= -28 \text{ kJ·mol}^{-1}$$

Comparing with the series in the text, magnesium oxide will be a weaker base than calcium oxide ($\Delta G° = -59$ kJ·mol^{-1}).

7.41 NO^+ is a Lewis acid, as it is an electron pair acceptor; similarly, Cl^- is a Lewis base, as it is an electron pair donor (when they combine to form NOCl).
$(NO)(AlCl_4)(NOCl) + [(CH_3)_4N]Cl(NOCl) \Rightarrow [(CH_3)_4N](AlCl_4)(NOCl) + NOCl(l)$

7.43 $NH_3(NH_3) + NH_3(NH_3) \leftrightarrow NH_4^+(NH_3) + NH_2^-(NH_3)$

(a) $K = [NH_4^+][NH_2^-] = 1 \times 10^{-33}$ as $[NH_4^+] = [NH_2^-]$
$[NH_4^+] = 3 \times 10^{-17}$ mol·L^{-1}

(b) Let $[NH_4^+] = x$, then $[NH_2^-] = 1.0 + x$
$K = (x)(1.0 + x) = 1 \times 10^{-33}$
Assume $x \ll 1.0$
$x = 1 \times 10^{-33}$ mol·L^{-1} $= [NH_4^+]$

7.45 (a) No. The reactants have the combinations borderline-borderline and hard-hard, while the products have combinations borderline-hard and hard-borderline. These like combinations are preferred over the mixed combinations that would result in the products.

(b) Yes. For the reactants, hard titanium(IV) ion is combined with soft iodide ion, soft titanium(II) ion with hard fluoride ion. The products will be preferred where hard-hard and soft-soft combinations result.

7.47 (a) Greater than unity, as silver ion is soft, as is cyanide ion and chloride ion is hard.

(b) Less than unity, as both mercury and iodide ions are soft, chloride ion is hard.

7.49 (a) Because thallium(I) is a large soft cation, it will probably have an insoluble chloride, TlCl, and precipitate with the other soft cations, silver(I), lead(II), and mercury(I) in analysis group I.

(b) Rubidium ion would behave like the other alkali metals in analysis group V and not give a precipitate.

(c) Radium would probably behave like the other alkaline earth metals (analysis group IV) and give a precipitate of $RaCO_3$ with carbonate ion.

(d) Iron(III) would probably resemble chromium(III) and aluminum(III) and give a precipitate of $Fe(OH)_3$ upon addition of hydroxide ion in the analysis group III.

7.51 (a) $MgSO_4$, (b) CoS.

Beyond the Basics

7.53 We start by writing the relevant equilibria:

$H_2S(aq) + H_2O(l) \leftrightarrow H_3O^+(aq) + HS^-(aq)$ $\quad\quad K_{a1} = 8.9 \times 10^{-8}$

$HS^-(aq) + H_2O(l) \leftrightarrow H_3O^+(aq) + S^{2-}(aq)$ $\quad\quad K_{a2} = 1.2 \times 10^{-13}$

$CdS(s) \leftrightarrow Cd^{2+}(aq) + S^{2-}(aq)$ $\quad\quad K_{sp} = 1.6 \times 10^{-28}$

$FeS(s) \leftrightarrow Fe^{2+}(aq) + S^{2-}(aq)$ $\quad\quad K_{sp} = 6.3 \times 10^{-18}$

Then combine the acid ionization constants:

$$K_{a1} \times K_{a2} = \left(\frac{[H_3O^+][HS^-]}{[H_2S]}\right)\left(\frac{[H_3O^+][S^{2-}]}{[HS^-]}\right) = \left(\frac{[H_3O^+]^2[S^{2-}]}{[H_2S]}\right)$$

We can solve for sulfide ion concentration:

$$[S^{2-}] = (8.9 \times 10^{-8})(1.2 \times 10^{-13})\left(\frac{(0.010)}{(1.0)^2} \right) = 1.1 \times 10^{-22}$$

For cadmium ion: $[Cd^{2+}][S^{2-}] = (0.010)(1.1 \times 10^{-22}) = 1.1 \times 10^{-24}$.
This is greater than K_{sp}, thus cadmium sulfide will precipitate.

For iron(II) ion $[Fe^{2+}][S^{2-}] = 1.1 \times 10^{-24}$.
This is less than K_{sp}; thus iron(II) sulfide will not precipitate.

7.55   Mercury(II) ion is a soft acid, so it will be found as ores of soft bases, such as sulfide ion. Zinc is a borderline acid, so it can be found as ores of both hard and soft bases.

7.57 $H_2CO_3(aq) + MgSiO_4(s) \rightarrow H_2O(l) + SiO_2(s) + MgCO_3(s)$
Thus, over geological time, the atmospheric concentration of carbon dioxide has decreased, in part due to the formation of magnesium (and calcium) carbonate minerals.

7.59   Dimethylsulfoxide must be a softer base than water, thus favoring the soft acid copper(I) rather than the borderline copper(II) ion.

7.61   In terms of the HSAB concept, the harder calcium ion is likely to form a stronger bond to the water molecules of hydration than the softer barium ion, thus driving the reaction towards solution.
$CaCl_2(s) \rightarrow Ca^{2+}(aq) + 2\ Cl^-(aq)$
In thermodynamic terms, the enthalpy (and free energy) of hydration of the calcium ion compared to the lattice energy of calcium chloride must be greater than the enthalpy of hydration of the barium ion compared to the lattice energy of barium chloride.

Chapter 8

OXIDATION AND REDUCTION

Exercises

8.1 (a) Substance that will oxidize another, itself being reduced.

(b) Two-dimensional plot of free energy against temperature for series of reactions that involve elements and their oxides, sulfides, or chlorides, etc. An Ellingham diagram is used for the prediction of reaction feasibility.

8.3 (a) $4[N_{ox}(P)] + 6(-2) = 0$
$[N_{ox}(P)] = +3$

(b) $3(+1) + [N_{ox}(P)] + 4(-2) = 0$
$[N_{ox}(P)] = +5$

(c) $3(+1) + [N_{ox}(P)] = 0$
$[N_{ox}(P)] = -3$

(d) $[N_{ox}(P)] + 4(+1) = +1$
$[N_{ox}(P)] = -3$

(e) $[N_{ox}(P)] + 1(-2) + 3(-1) = 0$
$[N_{ox}(P)] = +5$

8.5 (a) -2; (b) $+2$; (c) -1; (d) $+6$; (e) -2.

The diagrams show how the answers for (a), (b), (c), and (e) were derived.

H $\overset{\cdot\cdot}{\underset{\cdot\cdot}{S}}$ H	$\overset{\cdot\cdot}{\underset{\cdot\cdot}{Cl}}$ $\overset{\cdot\cdot}{\underset{\cdot\cdot}{S}}$ $\overset{\cdot\cdot}{\underset{\cdot\cdot}{Cl}}$	H $\overset{\cdot\cdot}{\underset{\cdot\cdot}{S}}$ $\overset{\cdot\cdot}{\underset{\cdot\cdot}{S}}$ H	$\overset{\cdot\cdot}{\underset{\cdot\cdot}{O}}$ :: C :: $\overset{\cdot\cdot}{\underset{\cdot\cdot}{S}}$
+1 −2 +1	−1 +2 −1	+1 −1 −1 +1	−2 +4 −2
(a)	(b)	(c)	(e)

8.7 $-1, +1, +3, +5, +7$.

8.9 (a) $+1$, (b) $+2$, (c) $+3$, (d) $+4$, (e) $+5$. An increase by units of $+1$ from Group 13 to Group 17.

8.11 (a) Nickel from $+2$ to 0, carbon from 0 to $+2$.

(b) Manganese from $+7$ to $+2$, sulfur from $+4$ to $+6$.

8.13　$NH_4^+(aq) + 3 H_2O(l) \rightarrow NO_3^-(aq) + 10 H^+(aq) + 8 e^-$

8.15　$N_2H_4(aq) + 4 OH^-(aq) \rightarrow N_2(g) + 4 H_2O(l) + 4 e^-$

8.17　(a)　$5 HBr(aq) + HBrO_3(aq) \rightarrow 3 Br_2(aq) + 3 H_2O(l)$

　　　(b)　$2 HNO_3(aq) + Cu(s) + 2 H^+(aq) \rightarrow$
$$2 NO_2(g) + Cu^{2+}(aq) + 2 H_2O(l)$$

8.19　(a)　$12 V(s) + 10 ClO_3^-(aq) + 18 OH^-(aq) \rightarrow$
$$6 HV_2O_7^{3-}(aq) + 10 Cl^-(aq) + 6 H_2O(l)$$

　　　(b)　$2 S_2O_4^{2-}(aq) + 3 O_2(g) + 4 OH^-(aq) \rightarrow$
$$4 SO_4^{2-}(aq) + 2 H_2O(l)$$

8.21　(a)　$Cu^+(aq) \rightarrow Cu^{2+}(aq) + e^-$　　　　$E° = -0.16$ V
　　　　　$Cu^+(aq) + e^- \rightarrow Cu(s)$　　　　　$E° = +0.52$ V
　　　Net E° is positive hence spontaneous.
　　　(b)　$Fe^{2+}(aq) \rightarrow Fe^{3+}(aq) + e^-$　　　　$E° = -0.77$ V
　　　　　$Fe^{2+}(aq) + 2 e^- \rightarrow Fe(s)$　　　　$E° = -0.44$ V
　　　Net E° is negative hence nonspontaneous.

8.23　For $Cr^{3+} \rightarrow Cr^{2+}$, $E° = -0.424$ V.
　　　There are several possibilities for oxidizing agents, one being
　　　$Zn \rightarrow Zn^{2+}$, $E° = +0.762$ V.

8.25　(a) $Au^{3+}(aq) + 3 e^- \rightarrow Au(s)$ would be the stronger oxidizing agent.
　　　(b) $Al(s) \rightarrow Al^{3+}(aq) + 3 e^-$ would be the stronger reducing agent.

8.27　$Pb^{2+}(aq) + 2 e^- \rightarrow Pb(s)$　　　$E° = -0.125$ V

$$E = E° - \frac{RT}{nF} \ln\left(\frac{1}{[Pb^{2+}]}\right)$$

$$E = (-0.125 \text{ V}) - \frac{8.31 \text{ V} \cdot \text{C} \cdot \text{mol}^{-1} \cdot \text{K}^{-1} \times 298 \text{ K}}{2 \times (9.65 \times 10^4 \text{ C} \cdot \text{mol}^{-1})} \ln\left(\frac{1}{1.5 \times 10^{-5}}\right) = -0.267 \text{ V}$$

8.29   pH = 7.00, thus $[H^+] = 1.0 \times 10^{-7}$

$$E = E^\circ - \frac{RT}{nF} \ln\left(\frac{1}{[H^+]^4 (pO_2)}\right)$$

$$E = (+1.229 \text{ V}) - \frac{8.31 \text{ V} \cdot \text{C} \cdot \text{mol}^{-1} \cdot \text{K}^{-1} x298 \text{ K}}{4 \times (9.65 \times 10^4 \text{ C} \cdot \text{mol}^{-1})} \ln\left|\frac{1}{(1.0 \times 10^{-7})^4 \left(20 \text{ kPa}\middle/100 \text{ kPa}\right)}\right|$$

$$= +0.805 \text{ V}$$

8.31   (a) Br_2

(b) $BrO_3^-(aq) \rightarrow HBrO(aq)$ is a four-electron reduction (+5 to +1 oxidation state), so $\Delta G^\circ = -4(F)(+0.54)$.

$HBrO(aq) \rightarrow \frac{1}{2} Br_2(aq)$ is a one-electron reduction, so $\Delta G^\circ = -1(F)(+0.45)$.

Total $\Delta G^\circ = (-2.16F) + (-0.45F) = -2.61F$

$E^\circ = -(-2.61F)/5F = +0.52 \text{ V}$

(c) The reduction half-reaction does not contain H^+ or OH^-:

$$\frac{1}{2} Br_2(aq) + e^- \rightarrow Br^-(aq)$$

Hence the Nernst expression does not have a pH-dependent term.

8.33   The most thermodynamically stable oxidation state is +3. The metal is a very strong reducing agent while cerium(IV) is an oxidizing agent.

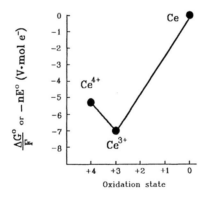

8.35   (a) Hg. (b) Mg. (c) Al. (d) Beyond this temperature, magnesium metal is a gas, hence three moles instead of one mole of gas are converted to two moles of solid. The decrease in entropy will be much greater and the slope will be steeper. (e) The hydrogen/water line has a positive slope. Thus at higher temperatures it will not cross many more metal/oxide lines than does carbon. At higher temperatures it will be no more effective as a reducing agent than it is at lower temperatures.

8.37   Consider a divalent metal M^{2+}. We can write an equation for oxide formation:

$M(s) + \frac{1}{2} O_2(g) \rightarrow MO(s)$

There will be a decrease in entropy, resulting from the loss of one-half mole of gas.

As $\Delta G = \Delta H - T\Delta S$, with ΔS being negative, increasing temperature will result in $(-T\Delta S)$ becoming more positive. Thus ΔG will become less negative with increasing temperature.

For carbon monoxide formation:

$C(s) + \frac{1}{2} O_2(g) \rightarrow CO(g)$

There is an increase of one-half moles of gas in the process, so ΔS will be positive. As a result, $(-T\Delta S)$ and ΔG will become more negative with increasing temperature.

8.39   Perchlorate ion is a stronger oxidizing agent at pH 0.00, as the reduction potential is +1.201 V at pH 0.00 compared to +0.374 V at pH 14.00.

Beyond the Basics

8.41 $C(s) + O_2(g) \rightarrow CO_2(g)$

$2 C(s) + O_2(g) \rightarrow 2 CO(g)$

The oxidation of carbon to carbon dioxide involves a near-zero entropy change; thus the slope of the ΔG versus temperature line will be close to zero. The oxidation to carbon monoxide, however, involves an increase of entropy (one mole of gas to two moles of gas); thus the $T\Delta S$ term will become increasingly negative with increase in temperature. The negative slope for this line will ultimately cross the carbon dioxide line, making the carbon monoxide production preferred at higher temperature.

For oxidation to carbon dioxide:

$\Delta H° = -394 \text{ kJ·mol}^{-1}$

$\Delta S° = [(+214) - (+205) - (+6)] \text{ J·mol}^{-1}\text{·K}^{-1} = +3 \text{ J·mol}^{-1}\text{·K}^{-1}$

$\Delta G° = (-394 \text{ kJ·mol}^{-1}) - T(+0.003 \text{ kJ·mol}^{-1}\text{·K}^{-1})$

For oxidation to carbon monoxide:

$\Delta H° = 2(-111) \text{ kJ·mol}^{-1} = -222 \text{ kJ·mol}^{-1}$

$\Delta S° = [2(+198) - (+205) - 2(+6)] \text{ J·mol}^{-1}\text{·K}^{-1} = +179 \text{ J·mol}^{-1}\text{·K}^{-1}$

$\Delta G^\circ = (-222 \text{ kJ·mol}^{-1}) - T(-0.179 \text{ kJ·mol}^{-1}\cdot\text{K}^{-1})$

At the crossover point the values for ΔG° wil be the same, thus
$(-394 \text{ kJ·mol}^{-1}) - T(+0.003 \text{ kJ·mol}^{-1}\cdot\text{K}^{-1}) =$

$$(-222 \text{ kJ·mol}^{-1}) - T(+0.179 \text{ kJ·mol}^{-1}\cdot\text{K}^{-1})$$

$T(0.182 \text{ kJ·mol}^{-1}\cdot\text{K}^{-1}) = +172 \text{ kJ·mol}^{-1}$

$T = 766 \text{ K} = 493^\circ\text{C}$

8.43   In the well water at about pH 7 and E° about 0, the manganense will be present as the colorless 2+ ion. On exposure to air (oxidizing conditions), E° will increase to the point where insoluble (brown) manganese(III) oxide will be formed; thus discoloring the toilet bowl.

Chapter 9

PERIODIC TRENDS

Exercises

9.1 (a) A pair of elements in a compound whose sum of valence electrons adds up to eight.

(b) The relationship between an element and the element to its lower right in the Periodic Table. The diagonal relationship is found in the top left-hand corner of the Periodic Table.

9.3 An alum has the general formula: $M^+M^{3+}(SO_4^{2-})_2 \cdot 12H_2O$ or more correctly, $M^+[M(OH_2)_6]^{3+}(SO_4^{2-})_2 \cdot 6H_2O$ where M^+ is a large monopositive ion such as potassium or ammonium and M^{3+} is a small tripositive ion such as aluminum, chromium(III), or iron(III).

9.5 (a) The melting point decreases down the Group as a result of the metallic bond throughout the metal lattice being weaker as the metal atoms become larger as the Group is descended.

(b) The melting point increases down the Group as the dispersion (London) forces between pairs of the covalently bonded atoms becomes stronger due to the increasing number of electrons.

(c) Carbon, at the top of the Group 14 elements, is network covalently bonded and thus has a very high melting point. So is silicon. With germanium, tin, and lead, there is weak metallic bonding and resulting low melting points.

9.7 KH (ionic); CaH_2 (ionic); GaH_3 (ionic); GeH_4 (network covalent); AsH_3 (small molecule covalent); H_2Se (small molecule covalent); HBr (small molecule covalent). This assumes the diagonality (Table 9.11) persists into Period 4. In fact, GaH_3 is polymeric covalent while GeH_4, like SiH_4, is a small molecule covalent compound. Thus, although five of the seven predictions are correct, those on the bonding borderline are incorrect. Trends in inorganic chemistry are rarely as precise as we would like them to be.

9.9 (a) They both form strongly oxidizing oxo-anions, MnO_4^- and ClO_4^-.
They both form explosive oxides, Mn_2O_7 and Cl_2O_7.
However, chlorine(VII) compounds are mostly colorless while those of manganese(VII) are intensely colored.
(b) There are in fact almost no similarities of these two ions.
Few silver(I) salts are soluble while nearly all rubidium salts are soluble.
Silver(I) forms a normal oxide, Ag_2O, while rubidium forms the dioxide(2–), RbO_2.

9.11 (a) A fused double ring structure, $B_5N_5H_8$, with alternating boron and nitrogen atoms.
(b) An end-on linked structure of two six-membered rings, $B_6N_6H_{10}$, with alternating boron and nitrogen atoms and having the boron of one ring linked to the nitrogen of the other six-membered ring.

9.13 This would have the structure $N \equiv C - C \equiv N$. It is probable that the π-bond system extends across the whole molecule, increasing the bond order across the C–C link while slightly weakening the $C \equiv N$ bond.

9.15 The ammonium ion is large and hence resembles the lower alkali metal ions. In particular, its non-spherical shape makes it resemble the largest of the alakli metal ions.

9.17 (a) Hydrocyanic acid is a weak acid like hydrofluoric acid.
(b) Silver cyanide reacts with ammonia to give the soluble diamminesilver ion as does the silver chloride, or, cyanogen reacts with water to give the cyanide and cyanate ions in a parallel reaction to that of chlorine, or, cynaide forms complex such as $[Cu(CN)_4]^{2-}$ which parallel their chloride counterparts.
(c) Cyanide ion is oxidized by copper(II) ion to give cyanogen, just as iodide ion is oxidized to iodine.

9.19 $SO_3(s) + H_2O(l) \rightarrow H_2SO_4(aq)$
$CrO_3(s) + H_2O(l) \rightarrow H_2CrO_4(aq)$

9.21   (a) Al_2O_3, Sc_2O_3.

(b) P_2O_5, V_2O_5 (actually the phosphorus(V) oxide is usually written as P_4O_{10} to reflect its molecular structure).

9.23   Tin.

9.25   The other elements exist as dimers, N_2, O_2, F_2, thus they have nearly twice the number of electrons as monatomic neon and hence significantly higher melting points resulting from the stronger dispersion (London) forces.

9.27   Europium has the electron configuration: $[Xe]6s^2 4f^7$. Forming the Eu^{2+} ion would retain the half-filled d orbital set.

9.29   The higher 3rd ionization energy of europium compared with that of the other lanthanoids means that the formation of EuX_3 compounds is thermodynamically less favorable than that of the other trivalent lanthanoids (a Born-Haber cycle will show this). As a result, the formation of EuX_2 compounds becomes more feasible.

9.31   (a) Copper(I) and indium(I); (b) cadmium(II) and lead(II).

9.33   Thallium(I) bromide.

9.35   (a) $C{\equiv}O$; (b) $(C{\equiv}C)^{2-}$.

9.37   Yttrium, as its 3+ ion will be closest in size to the lanthanoid ions. In fact, yttrium was discovered along with many of the lanthanoid elements in ores near the Swedish town of Ytterby. Hence the name of yttrium and those of ytterbium, terbium, and erbium.

Beyond the Basics

9.39 $SiH_4(g) + 2\ O_2(g) \rightarrow SiO_2(s) + 2\ H_2O(l)$

9.41   They would then be linked by the knight's move relationship.

9.43   Cesium-137 is a particularly hazardous material as all common cesium salts are highly water soluble, thus any water, such as rain, on a broken gauge will result in the radioactive cesium dispersing into the environment.

9.45 IF_7. The compound does exist. The non-existence of ClF_7 might result from the central chlorine atom being too small for seven large iodine atoms to be fitted around it.

9.47   +1, +2, +3, −4, −3, −2, −1 for Period 2.
       +1, +2, +3, +4, ±3, −2, −1 for Period 3.
       Though the signs change at different places in the two Periods, there is a trend to a maximum value of four followed by a decrease back down to one.

Chapter 10

HYDROGEN

Exercises

10.1   (a) Hydrogen atom bridging two (or more) other atoms in a covalent bond in which the hydrogen is less electronegative than the atoms to which it is bonded.
(b) Hydrogen atom bridging two (or more) other atoms in a covalent bond in which the hydrogen is more electronegative than the atoms to which it is bonded.

10.3   The ice cube consists of "heavy" water, deuterium oxide, which has a higher density than normal water and will therefore sink in normal water.

10.5   The difference in absorption frequency is very small, about 10^{-6} of the signal itself. Thus ppm is a convenient unit.

10.7   Unlike the halogens, hydrogen rarely forms a negative ion, nor is it highly reactive like the halogens.

10.9   Enthalpy driven. The chemical equation is
$$N_2(g) + 3\ H_2(g) \rightarrow 2\ NH_3(g)$$
there is a decrease in the number of gas molecules, hence a decrease in entropy. Thus for the reaction to be spontaneous, the reaction has to be exothermic.

10.11   (a) $2\ KHCO_3(s) \rightarrow K_2CO_3(s) + H_2O(g) + CO_2(g)$
(b) $HC{\equiv}CH(g) + 2\ H_2(g) \rightarrow H_3C{-}CH_3(g)$
(c) $PbO_2(s) + 2\ H_2(g) \rightarrow Pb(s) + 2\ H_2O(g)$
(d) $CaH_2(s) + H_2O(l) \rightarrow Ca(OH)_2(aq) + H_2(g)$

10.13   See the diagram. The much lesser enthalpy of formation of ammonia compared to water can be explained in terms of the much greater bond energy of dinitrogen (945 kJ·mol^{-1}) compared with that of dioxygen (498 kJ·mol^{-1}).

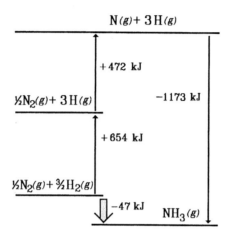

10.15   There are three categories of covalent hydrides: those in which the hydrogen is nearly neutral; those in which it is quite positive, and those in which it is negative. Most covalent hydrides belong in the first category; these are the low-polarity hydrides with low boiling points (except the long-chain hydrocarbons, where the dispersion forces become very significant). The second category consists of ammonia, water, and hydrogen fluoride, where hydrogen bonding (a protonic bridge) raises the melting and boiling points above those of the first category. Finally, there are the electron-deficient compounds of boron, where hydridic bridges form part of the molecular structure.

10.17   KH, CaH$_2$, GaH$_3$, GeH$_4$, AsH$_3$, H$_2$Se, HBr. The trend is increasing by one H until germanium, then a stepwise decrease by one H to hydrogen bromide. The first two members of the series are ionic, the remainder covalently bonded.

10.19   The ability to hydrogen bond and the low polarity of the carbon-hydrogen bond.

10.21   The closeness of the electronegativities of hydrogen and carbon, and the ability to hydrogen bond.

Beyond the Basics

10.23   (a) Yes, liquid; (b) no, gas; (c) yes, liquid; (d) no, gas.

10.25 We can calculate the B–H bridging bond energy by means of an energy diagram. The bond formation can be divided into two steps: the formation of four B–H terminal bonds and the formation of four B–H bridging bonds (see below). When we do this, a value of 208 kJ·mol^{-1} per mole of B–H bridging bonds is obtained almost exactly half of the 389 kJ·mol^{-1} of a normal B–H bond. This fits well with the concept of a single bonding pair shared by the two bonds.

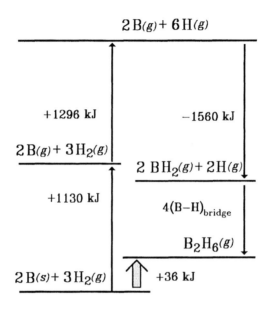

10.27 Energy per gram diborane = (2165 kJ·mol^{-1})/(16.8 g·mol^{-1}) = 129 kJ·g^{-1}

Energy per gram ethane = (1560 kJ·mol^{-1})/(30.0 g·mol^{-1}) = 52.0 kJ·g^{-1}

$B_2H_6(g) + 3\ O_2(g) \rightarrow B_2O_3(s) + 3\ H_2O(l)$

$C_2H_6(g) + {}^7/_2\ O_2(g) \rightarrow 2\ CO_2(g) + 3\ H_2O(l)$

ΔS for diborane combustion = [(+54) + 3(+70) − (+232) − 3(+205)] J·mol^{-1}·K^{-1}

     = −583 J·mol^{-1}·K^{-1}

ΔS for ethane combustion

     = [2(+214) + 3(+70) − (+230) − ${}^7/_2$(+205)] kJ·mol^{-1}·K^{-1}

     = −310 J·mol^{-1}·K^{-1}

The difference in values is a result of the boron oxide being a solid while the carbon oxide is a gas (much higher entropy). The practical diasadvantages are the cost of the diborane, the difficulty in handling such a flammable fuel, and the clogging of the engine by the solid boron oxide product.

10.29 Hydrogen and carbon monoxide.

$$H_2O(l) + C(s) \rightarrow H_2(g) + CO(g)$$

The combustion reaction would therefore be:

$$H_2(g) + CO(g) + O_2(g) \rightarrow H_2O(g) + CO_2(g)$$

$\Delta H = [(1\ \text{mol})(-242) + (1\ \text{mol})(-394) - (1\ \text{mol})(-111)]\ \text{kJ·mol}^{-1} = -525\ \text{kJ}$

Per mole, this is $-525/2\ \text{kJ·mol}^{-1} = -262\ \text{kJ·mol}^{-1}$, compared with $-242\ \text{kJ·mol}^{-1}$ for the combustion of pure dihydrogen. So there is about an 8 percent higher energy availability provided that we consider the water to be produced in the gas phase (assuming the synthesis does produce the precise stoichiometric mixture).

Chapter 11

THE GROUP 1 ELEMENTS: THE ALKALI METALS

Exercises

11.1   (a) $2\ Na(s) + 2\ H_2O(l) \rightarrow 2\ NaOH(aq) + H_2(g)$

      (b) $Rb(s) + O_2(g) \rightarrow RbO_2(s)$

      (c) $2\ KOH(s) + CO_2(g) \rightarrow K_2CO_3(s) + H_2O(l)$

      (d) $2\ NaNO_3(s) \rightarrow 2\ NaNO_2(s) + O_2(g)$

11.3   They resemble "typical" metals in that they are shiny and silvery and good conductors of heat and electricity. They differ from "typical" metals in that they are soft, extremely chemically reactive, and have low melting points and very low densities.

11.5   Any three of the following:

All common chemical compounds are water-soluble.

They always form ions of +1 oxidation state.

Their compounds are almost always ionic.

The low charge density alkali metal ions stabilize large low-charge anions such as hydrogen carbonate.

Their compounds are rarely hydrated.

11.7   The most likely argument is that the hydroxide ion can hydrogen bond with the surrounding water molecules whereas the chloride ion can only form ion-dipole attractions. This hydrogen bond formation would release more energy than the ion-dipole formation. Alternatively, we can simply argue that the smaller hydroxide ion will have a higher charge density and the ion-dipole attractions will be greater—this is the same argument, just a different approach. In fact, the hydration enthalpy for the hydroxide ion is indeed much greater than that of the chloride ion.

11.9   Because the equilibrium of the synthesis reaction

$$Na(l) + KCl(l) \leftrightarrow K(l) + NaCl(l)$$

lies to the left. To make the reaction shift right, the potassium, which boils at a lower temperature than sodium, must be continuously removed as a gas.

11.11 (a) Sodium hydroxide; (b) anhydrous sodium carbonate; (c) sodium carbonate decahydrate.

11.13 (a) Loss of water by a hydrated salt in a low humidity environment.
(b) Chemical similarities of one element and the element to its lower right in the periodic table.

11.15
$$CO_2(g) + NH_3(aq) + H_2O(l) \rightarrow NH_4^+(aq) + HCO_3^-(aq)$$
$$HCO_3^-(aq) + Na^+(aq) \rightarrow NaHCO_3(s)$$
$$2\,NaHCO_3(s) \rightarrow Na_2CO_3(s) + H_2O(g) + CO_2(g)$$
$$CaCO_3(s) \rightarrow CaO(s) + CO_2(g)$$
$$CaO(s) + H_2O(l) \rightarrow Ca(OH)_2(s)$$
$$2\,NH_4^+(aq) + 2\,Cl^-(aq) + Ca(OH)_2(s)$$
$$\rightarrow 2\,NH_3(aq) + CaCl_2(aq) + 2\,H_2O(l)$$

The problems are the disposal of waste calcium chloride and the high energy requirements of the process.

11.17 The ammonium ion is monopositive, like the alkali metals (unlike most other metals that have 2+ or higher charge); its salts are all soluble like those of the alkali metal salts; its size is about the middle of the alkali metal ion range; all its common salts are colorless, like those of the alkali metals.

11.19 Potassium dioxide(1−) has a lower molar mass than the equivalent cesium compound. For launch, minimum mass for the same oxygen-generating capacity is crucial. Also, the common potassium salts are much lower in cost than those of cesium.

11.21 Lithium:

$6 \text{ Li}(s) + \text{N}_2(g) \rightarrow 2 \text{ Li}_3\text{N}(s)$

$2 \text{ Li}(s) + \text{Cl}_2(g) \rightarrow 2 \text{ LiCl}(s)$

$\text{Li}(s) + \text{C}_4\text{H}_9\text{Cl}(solv) \rightarrow \text{LiC}_4\text{H}_9(solv) + \text{LiCl}(s)$

$4 \text{ Li}(s) + \text{O}_2(g) \rightarrow 2 \text{ Li}_2\text{O}(s)$

$2 \text{ Li}(s) + \text{H}_2\text{O}(l) \rightarrow 2 \text{ LiOH}(aq) + \text{H}_2(g)$

$\text{Li}_2\text{O}(s) + \text{H}_2\text{O}(l) \rightarrow 2 \text{ LiOH}(aq)$

$2 \text{ LiOH}(aq) + \text{CO}_2(g) \rightarrow \text{Li}_2\text{CO}_3(aq) + \text{H}_2\text{O}(l)$

$\text{Li}_2\text{O}(s) + \text{CO}_2(g) \rightarrow \text{Li}_2\text{CO}_3(s)$

Sodium:

$2 \text{ Na}(s) + \text{Cl}_2(g) \rightarrow 2 \text{ NaCl}(s)$

$2 \text{ Na}(s) + \text{H}_2\text{O}(l) \rightarrow 2 \text{ NaOH}(aq) + \text{H}_2(g)$

$2 \text{ Na}(s) + \text{O}_2(g) \rightarrow \text{Na}_2\text{O}_2(s)$

$\text{Na}_2\text{O}_2(g) + \text{H}_2\text{O}(l) \rightarrow 2 \text{ NaOH}(aq) + \text{H}_2\text{O}_2(aq)$

$2 \text{ NaOH}(aq) + \text{CO}_2(g) \rightarrow \text{Na}_2\text{CO}_3(aq) + \text{H}_2\text{O}(l)$

$\text{Na}_2\text{CO}_3(aq) + \text{CO}_2(g) + \text{H}_2\text{O}(l) \rightarrow 2 \text{ NaHCO}_3(aq)$

$2 \text{ Na}(s) + 2 \text{ NH}_3(l) \rightarrow 2 \text{ NaNH}_2(NH_3) + \text{H}_2(g)$

$\text{Na}_2\text{O}_2(s) + \text{CO}_2(g) \rightarrow \text{Na}_2\text{CO}_3(s) + \text{O}_2(g)$

Potassium

$\text{Na}(l) + \text{KCl}(l) \rightarrow \text{K}(g) + \text{NaCl}(l)$

$2 \text{ K}(s) + \text{Cl}_2(g) \rightarrow 2 \text{ KCl}(s)$

$\text{K}(s) + \text{O}_2(g) \rightarrow \text{KO}_2(s)$

$2 \text{ KO}_2(s) + 2 \text{ H}_2\text{O}(l) \rightarrow 2 \text{ KOH}(aq) + \text{H}_2\text{O}_2(aq) + \text{O}_2(g)$

$2 \text{ K}(s) + 2 \text{ H}_2\text{O}(l) \rightarrow 2 \text{ KOH}(aq) + \text{H}_2(g)$

$2 \text{ KOH}(aq) + \text{CO}_2(g) \rightarrow \text{K}_2\text{CO}_3(aq) + \text{H}_2\text{O}(l)$

$\text{K}_2\text{CO}_3(aq) + \text{CO}_2(g) + \text{H}_2\text{O}(l) \rightarrow 2 \text{ KHCO}_3(aq)$

$2 \text{ KO}_2(s) + \text{CO}_2(g) \rightarrow \text{K}_2\text{CO}_3(s) + 2 \text{ O}_2(g)$

$3 \text{ K}^+(aq) + [\text{Co(NO}_2)_6]^{3-}(aq) \rightarrow \text{K}_3[\text{Co(NO}_2)_6](s)$

Beyond the Basics

11.23   Mass of sodium $= 1.00 \text{ t} \times (10^6 \text{ g})/(1 \text{ t}) = 1.00 \times 10^6 \text{ g}$

Mole sodium $= (1.00 \times 10^6 \text{ g}) \times (1 \text{ mol})/(23.0 \text{ g}) = 4.35 \times 10^4 \text{ mol}$

$\text{Na}^+(NaCl) + \text{e}^- \rightarrow \text{Na}(l)$

Thus mole electrons $= 4.35 \times 10^4 \text{ mol}$

$$\text{Current} = (4.35 \times 10^4 \text{ mol}) \times \frac{(9.65 \times 10^4 \text{ A} \cdot \text{V} \cdot \text{s})}{(1 \text{ mol})(7.0 \text{ V})(86400 \text{ s})} = 6.94 \times 10^4 \text{ A}$$

11.25   In the series LiF to CsF, there is an increasing mismatch in ion sizes; thus the lattice energy will decrease more than otherwise expected. As a result, the enthalpies of formation will decrease. For the series LiI to CsI, there is a decreasing mismatch in ion sizes; thus the lattice energy will decrease to a lesser extent than otherwise expected. As a result the enthalpies of formation will increase.

11.27   Sodium fluoride will be the less soluble, as there is a close match in ion sizes resulting in a higher lattice energy ($\text{Na}^+ = 116$ pm; $\text{F}^- = 117$ pm). There is a mismatch in sizes with the tetrafluoroborate ion (which actually has a radius of 218 pm). Thus the hydration energy will more probably exceed the (lower) lattice energy, making the compound more soluble.

11.29   There are two possible answers: that there is appreciable covalent bonding in the lithium hydride, thus reducing the Li–H separation, or that the lithium ion is so small that the lattice consists of touching hydride ions with lithium ions "rattling around" in the lattice holes—this is certainly true of lithium iodide.

11.31   LiF and KI.

11.33   Calcium-40 is a "doubly-magic" nucleus with filled shells of protons and neutrons.

Chapter 12

THE GROUP 2 ELEMENTS: THE ALKALINE EARTH METALS

Exercises

12.1   (a) $2\ Ca(s) + O_2(g) \rightarrow 2\ CaO(s)$

        (b) $CaCO_3(s) \rightarrow CaO(s) + CO_2(g)$

        (c) $Ca(HCO_3)_2(aq) \rightarrow CaCO_3(s) + H_2O(l) + CO_2(g)$

        (d) $CaO(s) + 3\ C(s) \rightarrow CaC_2(s) + CO(g)$

12.3   (a) Barium; (b) barium.

12.5   Even though there will be a greater increase in entropy when the magnesium chloride lattice vaporizes compared to that of the sodium chloride lattice (three moles of ions instead of two), the higher charge density magnesium ion will cause the water molecules surrounding it during the hydration step to become much more ordered than with the lower charge density sodium ion. It is this significant difference in hydration entropy that causes the considerable difference in solution entropy change.

12.7   They form 2+ ions exclusively and their salts tend to be highly hydrated.

12.9   The commonly accepted explanation is that the beryllium 2+ cation is so small that six water molecules cannot fit around it.

12.11   Rainwater, an aqueous solution of carbon dioxide, percolates into limestone deposits, reacting with the calcium carbonate to give a solution of calcium hydrogen carbonate. The solution flows away, leaving a space that, with continued rain, enlarges to cave size.

$$CaCO_3(s) + CO_2(aq) + H_2O(l) \rightarrow Ca(HCO_3)_2(aq)$$

12.13   Calcium hydroxide is added to seawater causing precipitation of the less soluble magnesium hydroxide:

$$Ca(OH)_2(aq) + Mg^{2+}(aq) \rightarrow Mg(OH)_2(s) + Ca^{2+}(aq)$$

The magnesium hydroxide is filtered off and neutralized with hydrochloric acid:

$$Mg(OH)_2(s) + 2\ HCl(aq) \rightarrow MgCl_2(aq) + 2\ H_2O(l)$$

The solution is evaporated to dryness and electrolysed:

$$Mg^{2+}(MgCl_2) + 2\ e^- \rightarrow Mg(l)$$
$$2\ Cl^-(MgCl_2) \rightarrow Cl_2(g) + 2\ e^-$$

(It can be added that the chlorine is recycled to produce the hydrochloric acid reagent, while the calcium hydroxide comes from limestone by heating to calcium oxide and adding water.)

12.15   (a) $Ca(OH)_2$ (hydrated lime) or CaO (quicklime); (b) $Mg(OH)_2$;
        (c) $MgSO_4 \cdot 7H_2O$.

12.17   The absorption of X-rays depends upon the square of the atomic number of the element. Thus lead is used because it has the highest atomic number of the common, non-radioactive elements (and it is a soft, malleable metal, so it is feasible to make lead aprons and other shields).

12.19   Both metals form tough oxide coatings over their surface that protect the bulk of the metal from oxidation; the two metals are amphoteric, forming beryllate and aluminate anions; they both form carbides containing the C^{4-} ion.

12.21   Magnesium ion is a key component of chlorophyll, the one molecule that has been crucial to converting our planet's atmosphere from carbon dioxide to dioxygen.

12.23   (a) $Mg(s) + HCl(aq) \rightarrow MgCl_2(aq) + H_2(g)$
               then evaporate to dryness to crystallize $MgCl_2 \cdot 6H_2O(s)$.
      (b) $Mg(s) + Cl_2(g) \rightarrow MgCl_2(s)$

Beyond the Basics

12.25 $CaSO_4 \cdot 2H_2O(s) \rightarrow CaSO_4 \cdot \frac{1}{2}H_2O(s) + 1\frac{1}{2}\ H_2O(g)$

$$\Delta H° = [1(-1577) + 1\frac{1}{2}(-242) - 1(-2023)]\ kJ \cdot mol^{-1}$$
$$= +83\ kJ \cdot mol^{-1}$$
$$\Delta S° = [1(131) + 1\frac{1}{2}(189) - 1(194)]\ J \cdot mol^{-1} \cdot K^{-1}$$
$$= +220\ J \cdot mol^{-1} \cdot K^{-1} = 0.220\ kJ \cdot mol^{-1} \cdot K^{-1}$$
$$\Delta G° = (+83\ kJ \cdot mol^{-1}) - T(+0.220\ kJ \cdot mol^{-1} \cdot K^{-1}) = 0$$
$$T = 377\ K = 104°C \text{ (close to the actual value of about } 100°C)$$

(If liquid water is used as product, a temperature of 132°C is calculated, which indicates that the value for $H_2O(g)$ has to be used.)

12.27 Except for the second period, most metal ions are hydrated by six water molecules. Thus the formula is actually $[Mg(OH_2)_6]^{2+}[SO_4 \cdot H_2O]^{2-}$. It is also possible to use the analogy with the zinc sulfate heptahydrate formula (Chapter 5).

12.29 BeH^+, as the molecular orbital diagrams show. This ion would possess a single bond, while the neutral molecule would have a ½ bond and BeH^- would not exist.

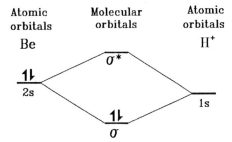

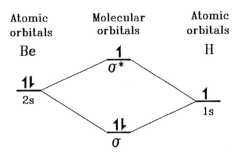

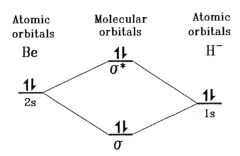

12.31 $Ca_3N_2(s) + 4 NH_3(l) \rightarrow 3 Ca(NH_2)_2(NH_3)$

12.33 $BeF_2(s) + Mg(s) \rightarrow MgF_2(s) + Be(s)$

$\Delta G^\circ = [(-1071) - (-979)]$ kJ·mol^{-1} = -92 kJ·mol^{-1}

Less favorable, for at a higher temperature, the low-melting magnesium will be a liquid:

$BeF_2(s) + Mg(l) \rightarrow MgF_2(s) + Be(s)$

As liquids have higher entropies than solids, there will be a decrease in entropy for the process, and hence as $\Delta G = \Delta H - T\Delta S$, a less-negative free energy change. The reason for synthesizing at a higher temperature is the greatly increased rate of reaction resulting from the liquid magnesium acting as a "solvent." At room temperature, reaction between two solids will be very slow.

12.35   The species is probably Na_2BeCl_4, containing sodium ions and the tetrahedral tetrachloroberyllate ion, $[BeCl_4]^{2-}$.

Chapter 13

THE GROUP 13 ELEMENTS

Exercises

13.1   (a) $3 K(l) + AlCl_3(s) \rightarrow Al(s) + 3 KCl(s)$

       (b) $B_2O_3(s) + 2 NH_3(g) \rightarrow 2 BN(s) + 3 H_2O(g)$

       (c) $2 Al(s) + 2 OH^-(aq) + 6 H_2O(l) \rightarrow 2 [Al(OH)_4]^-(aq) + 3 H_2(g)$

       (d) $2 B_4H_{10}(g) + 11 O_2(g) \rightarrow 4 B_2O_3(s) + 10 H_2O(g)$

13.3   In the hexahydrate ion, the Al^{3+} is surrounded by the partially negative oxygen atoms of the six water molecules with the partially positive hydrogens of the water molecules pointing away from the ion. In effect, then, the charge is delocalized over the whole hydrated ion rather than being concentrated in the small Al^{3+} "core."

13.5   See the energy cycle diagram. The major factors in the very large enthalpy of formation are the weak fluorine-fluorine bond, the weakest of the halogen bonds; and the exceedingly strong boron-fluorine bond (the latter can be explained either as a result of partial ionic character of the very polar B–F bond or some degree of π-orbital formation involving a full $2p$ orbital of the fluorine and an empty $2p$ orbital of the boron).

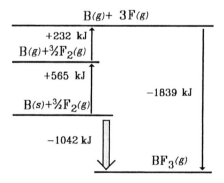

13.7 The surface layer of aluminum reacts to give aluminum oxide. The oxide ions fit into the lattice sites formerly occupied by the aluminum atoms, as they are about the same size. The aluminum ions are so small that they fit into interstices in the lattice. The aluminum oxide layer, as it has not disturbed the aluminum crystal structure, forms an impervious layer preventing attack of the metal beneath.

13.9 Aluminum is an amphoteric metal; that is, it is a weak—"borderline"—metal.

13.11 The potential environmental hazards are the "red mud"—the basic residue from the bauxite purification; hydrogen fluoride gas from hydrolysis of cryolite; the carbon oxides that are produced at the anode; and fluorocarbon compounds produced by reaction of fluorine with the carbon anode.

13.13 Aluminum fluoride is a typical ionic compound with a high melting point. Both aluminum bromide and aluminum iodide are covalently bonded dimers, Al_2Br_6 and Al_2I_6. Aluminum chloride is a borderline case where in the solid, it has an ionic structure which can alternatively be considered as network covalent, while in the liquid phase, it consists of covalently bonded dimers, Al_2Cl_6.

13.15 A spinel has the formula AB_2X_4, where A is a dipositive metal ion, B is a tripositive metal ion, and X is a dinegative ion. The negative ion forms a cubic close-packed arrangement, and in the normal spinel, the A cations occupy tetrahedral sites and the B cations octahedral sites. In the reverse spinel, the A cations occupy octahedral sites while half of the B cations occupy the tetrahedral sites, the other half octahedral sites.

13.17 Gallium(III) fluoride must consist of an ionic lattice of gallium(3+) and chloride(1−) ions while gallium(III) chloride must contain discrete $GaCl_3$ molecules.

13.19 Aluminum ion forms the highly insoluble aluminum hydroxide in neutral conditions. However, in acid conditions, such as lakes acidified by acid rain, the soluble $Al(OH_2)_6^{3+}$ is produced. The aluminum ion is very toxic to fish.

Beyond the Basics

13.21 The metallic radius is a measure of the atomic size when packed in a metal crystal lattice. It will not be much smaller than the van der Waals (non-bonding) radius. The covalent radius will be smaller as there is orbital overlap with the atom to which it is covalently bonded. The ionic radius is by far the smallest as all the valence electrons have been lost and the measure is simply of the core of the ion.

13.23 $Cl_3Al[O(C_2H_5)_2]$, where a lone pair on the oxygen of the ether occupies the fourth bonding site around the aluminum ion.

13.25 As we might expect from the diagonal relationship (Chapter 9), the beryllium ion will resemble the aluminum ion. Hence:
$[Be(OH_2)_4]^{2+}(aq) + H_2O(l) \leftrightarrow [Be(OH_2)_3(OH)]^+(aq) + H_3O^+(aq)$

13.27 Let number of ions of magnesium $= x$, then:
$1 \times (K^+) + x \times (Mg^{2+}) + 1 \times (Al^{3+}) + 3 \times (Si^{4+}) + 10 \times (O^{2-}) + 2 \times (OH^-) = 0$
$1(+1) + x(+2) + 1(+3) + 3(+4) + 10(-2) + 2(-1) = 0$
$x = +3$.

13.29 Setting up the disproportionation reaction:
$3\ GaCl(s) \rightarrow GaCl_3(s) + 2\ Ga(s)$
$\Delta H = \Sigma\Delta H_f°(products) - \Sigma\Delta H_f°(reactants)$
$\quad = [1(-525) + 2(0)] - [1(+38)] = -563\ kJ \cdot mol{-}1$
as there are equal moles (in the same phase) on each side of the equation, the free energy for the disproportionation should be equally high. Thus disproportionation should occur.

13.31 $4\ AlCl_3(s) + CH_3CN(l) \rightarrow [Al(CH_3CN)_6]^{3+}(CH_3CN) + 3\ [AlCl_4]^-(CH_3CN)$

13.33 $Ga(OH_2)_6{}^{3+}(aq) \rightarrow GaO(OH)(s) + H_2O(l) + 3\ H_3O^+(aq)$
Addition of acid (hydronium ion) will shift the equilibrium to the left.

13.35 As gallium exhibits only oxidation states of $+1$ and $+3$, the compound is probably $[Ga^+][GaCl_4]^-$, the gallium in the anion being in the $+3$ oxidation state.

13.37 $B(OH)_3(aq) + H_2O(l) \rightarrow B(OH)_4^-(aq) + H^+(aq)$

This can also be written as:

$H_3BO_3(aq) + H_2O(l) \rightarrow B(OH)_4^-(aq) + H^+(aq)$

13.39

$$\left[\ddot{\text{Ö}}::\text{B}::\ddot{\text{Ö}}\right]^- \quad \left[\ddot{\text{C}}::\text{B}::\ddot{\text{C}}\right]^{5-} \quad \left[\ddot{\text{N}}::\text{B}::\ddot{\text{N}}\right]^{3-}$$

13.41 For sodium chloride packing, the radius ratio, r_+/r_-, must be between 0.414 and 0.732.

From the icosahedral structure of B_{12} (Figure 13.1), the radius of the boron cluster would seem to be about the sum of two atoms, 176 pm. If we assume the zirconium(IV) ion, this would give a radius ratio of about 0.41, essentially what could be expected to form an NaCl lattice. Using the atomic radius of zirconium would give a ratio of sizes of close to unity. Not what one would expect for an NaCl packing pattern. The structure must be $[Zr^{4+}][B_{12}^{4-}]$.

Chapter 14

THE GROUP 14 ELEMENTS

Exercises

14.1   (a) $Li_2C_2(s) + 2\ H_2O(l) \rightarrow 2\ LiOH(aq) + C_2H_2(g)$

      (b) $SiO_2(s) + 2\ C(s) \rightarrow Si(l) + 2\ CO(g)$

      (c) $CuO(s) + CO(g) \rightarrow Cu(s) + CO_2(g)$

      (d) $Ca(OH)_2(aq) + CO_2(g) \rightarrow CaCO_3(s) + H_2O(l)$

              $CaCO_3(s) + CO_2(g) + H_2O(l) \rightarrow Ca(HCO_3)_2(aq)$

      (e) $CH_4(g) + 4\ S(l) \rightarrow CS_2(g) + 2\ H_2S(g)$

      (f) $SiO_2(s) + 2\ Na_2CO_3(l) \rightarrow Na_4SiO_4(s) + 2\ CO_2(g)$

      (g) $PbO_2(s) + 4\ HCl(aq) \rightarrow PbCl_4(aq) + 2\ H_2O(l)$

              $PbCl_4(aq) \rightarrow PbCl_2(s) + Cl_2(g)$

14.3   (a) Ability of an element to form chains of its atoms.

     (b) Silicates in which there are numerous cavities in the structure resulting in very low densities.

     (c) Non-metallic inorganic compounds prepared by high temperature synthesis.

     (d) Chains of alternating silicon and oxygen atoms to which organic (carbon-containing) side groups are attached.

14.5   Diamond is a very hard, transparent, colorless solid that is a good conductor of heat but a non-conductor of electricity. It is insoluble in all solvents and it is chemically unreactive except upon heating in dioxygen. Graphite is a soft, slippery, black solid that is a poor conductor of heat but a good conductor of electricity. It is insoluble in all solvents and will react chemically only with very reactive elements such as dioxygen and difluorine. C_{60} is black and a nonconductor of heat and electricity. It is soluble in many non-polar and low-polarity solvents and it is quite reactive chemically.

14.7   Diamond and graphite both have network covalent bonded structures. The solvation process cannot provide the energy necessary to break non-polar covalent bonds. The fullerenes consist of discrete molecules, such as C_{60}. These individual non-polar units can become solvated by non-polar or low-polarity solvent molecules and hence dissolve.

14.9   The three classes are ionic, covalent, and metallic. Ionic carbides are formed by the most electropositive metals. These may contain the dicarbide(2–) ion, C_2^{2-} or the true carbide ion C^{4-}. Both types of ionic carbide react with water to produce the appropriate hydrocarbon. Covalent carbides are formed by nonmetals, specifically boron and silicon, more electronegative than carbon. These carbides are very hard and have high melting points. The metallic carbides are interstitial carbides, in that the carbon atoms fit in interstices within the metal structure. As such, they have many metallic properties, such as hardness, metallic luster, and electrical conductivity.

14.11   The chemical equation is
$$SiO_2(s) + 3\ C(s) \rightarrow SiC(s) + 2\ CO(g)$$
and with the net increase of two moles of gas, the reaction should be entropy driven. The need for high-temperature synthesis might indicate that there is not only a high activation energy but also the possibility of the reaction being endothermic (according to Le Chatelier's Principle).

$\Delta H° \quad = [1(-65) + 2(-111) - 1(-911) - 3(0)]\ kJ \cdot mol^{-1}$
$\qquad = +624\ kJ \cdot mol^{-1}$

$\Delta S° \quad = [1(+17) + 2(+198) - 1(+41) - 3(6)]\ J \cdot mol^{-1} \cdot K^{-1}$
$\qquad = +354\ J \cdot mol^{-1} \cdot K^{-1} = +0.354\ kJ \cdot mol^{-1} \cdot K^{-1}$

Thus there is indeed a substantial increase in entropy (the driving force for the reaction) while the reaction is also very endothermic.

$\Delta G° \quad = (+624\ kJ \cdot mol^{-1}) - (2273\ K)(+0.354\ kJ \cdot mol^{-1} \cdot K^{-1})$
$\qquad = -181\ kJ \cdot mol^{-1}$

At this temperature the reaction is spontaneous.

14.13 As can be seen from the enthalpy of the following formation diagrams, it is the lower bond energy of the C=S bond compared to the C=O that makes such a large difference to the enthalpy of formation values.

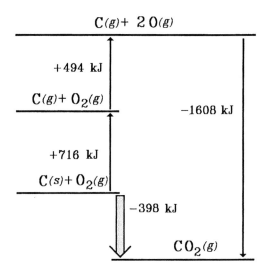

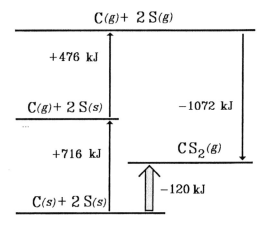

14.15 As the central carbon atom must form two σ bonds to the sulfur atoms, we assume that *sp* hybrid orbitals are formed. Like carbon dioxide, we consider that the 2*p* orbitals at right angles to the *sp* hybrid orbitals overlap with 3*p* orbitals of the sulfur to provide a pair of π bonds.

14.17 We explain the high activation energy of methane in terms of the lack of any empty orbitals on the carbon atoms that can be used for intermediate formation in the oxidation process. Silicon in the comparable silane molecule has empty $3d$ orbitals that can be involved in the oxidation process.

14.19 The synthesis of HFC-134a requires a complex, expensive multistep procedure. Also, higher pressures are necessary for its liquefaction, compared with the CFCs, making it necessary to completely replace the old refrigeration units when changing to HFC-134a.

14.21 Methane is a particularly potent greenhouse gas because it absorbs wavelengths in the infrared region that are currently transparent.

14.23 This ion is isoelectronic with nitrogen dioxide. It has a single unpaired electron, so the bond angle is greater than 120°. The bond order would be about 1½ compared to 2 in carbon dioxide.

$$\left[\; :\overset{..}{\underset{..}{O}}::C\overset{\cdot}{\underset{\cdot\cdot}{}}\overset{}{\underset{:\overset{..}{O}:}{}} \; \right]^{-}$$

14.25 Trigonal planar, as there appear to be three bonding directions around the central carbon. This would be one of three possible resonance structures resulting in an extended π-bond system throughout the ion. The average carbon-carbon bond order is $1\frac{1}{3}$.

$$\left[\; :N:::C:\overset{..}{\underset{..}{C}}:C:::N: \; \right]^{-}$$
$$\begin{array}{c} \overset{..}{\underset{..}{C}} \\ :N: \end{array}$$

14.27 As the Si_4O_{11} unit has a charge of 6−, for neutrality in the compound, the five iron ions must have a total charge of 12+. If the number of Fe^{2+} ions is x and the number of Fe^{3+} ions is y, we can say:

$$x(+2) + y(+3) = +12$$

but $(x+y) = 5$, the sum of the iron ions. Hence by substitution, $x = 3$ and $y = 2$. Thus there are three Fe^{2+} ions and two Fe^{3+} ions per formula.

14.29 Zeolites are used as ion exchangers for water; as adsorption agents, particularly for water in organic solvents; for gas separation, particularly dioxygen and dinitrogen from air; and most important, as specialized catalysts, particularly in the oil industry.

14.31 Silicone polymers are chemically very stable. Thus any polymer molecules that leak from the plastic sac in breast implants cannot be broken down by normal bodily processes.

14.33 Electron-dot structures of tin(IV) chloride and tin(II) chloride:

Corresponding molecular shapes (tetrahedral and vee-shaped) according to VSEPR theory.

14.35 The half-equations are:

$$PbO(s) + H_2O(l) \rightarrow PbO_2(s) + 2\ H^+(aq) + 2\ e^-$$
$$PbO(s) + 2\ H^+(aq) + 2\ e^- \rightarrow Pb(s) + H_2O(l)$$

14.37 CN^- and CO are the most common species isoelectronic with C_2^{2-}.

14.39 The major problem with inorganic polymer chemistry is the lack of the range of synthetic pathways compared with those available for organic polymer synthesis.

14.41 Carbon

$4\ CO(g) + Ni(s) \rightarrow Ni(CO)_4(g)$

$CO(g) + Cl_2(g) \rightarrow COCl_2(g)$

$CO(g) + S(s) \rightarrow COS(g)$

$HCOOH(l) \xrightarrow{H_2SO_4} CO(g) + H_2O(l)$

$CO_2(g) + 2\ Ca(s) \rightarrow C(s) + 2\ CaO(s)$

$2\ CO(g) + O_2(g) \rightarrow 2\ CO_2(g)$

$CO(g) + 2\ H_2(g) \xrightarrow{catalyst} CH_3OH(l)$

$2\ C(s) + O_2(g) \rightarrow 2\ CO(g)$

$C(s) + O_2(g) \rightarrow CO_2(g)$

$Na_2C_2(s) + 2\ H_2O(l) \rightarrow 2\ NaOH(aq) + C_2H_2(g)$

$2\ C_2H_2(g) + 5\ O_2(g) \rightarrow 4\ CO_2(g) + 2\ H_2O(l)$

$Al_4C_3(s) + H_2O(l) \rightarrow 3\ CH_4(g) + 4\ Al(OH)_3(s)$

$CH_4(g) + 2\ O_2(g) \rightarrow CO_2(g) + 2\ H_2O(l)$

$CH_4(g) + 4\ S(l) \rightarrow CS_2(g) + 2\ H_2S(g)$

$CS_2(g) + 3\ Cl_2(g) \rightarrow CCl_4(g) + S_2Cl_2(l)$

$CS_2(g) + S_2Cl_2(l) \rightarrow CCl_4(g) + 6\ S(s)$

$CH_4(g) + NH_3(g) \rightarrow HCN(g) + 3\ H_2(g)$

$HCN(aq) + H_2O(l) \rightarrow H_3O^+(aq) + CN^-(aq)$

$Ca(OH)_2(aq) + CO_2(g) \rightarrow CaCO_3(s) + H_2O(l)$

$CaCO_3(s) + 2\ HCl(aq) \rightarrow CaCl_2(aq) + H_2O(l) + CO_2(g)$

$CaCO_3(s) + H_2O(l) + CO_2(g) \rightarrow Ca(HCO_3)_2(aq)$

Silicon

$$Si(s) + HCl(g) \rightarrow SiHCl_3(g) + H_2(g)$$

$$2\ CH_3Cl(g) + Si(s) \rightarrow (CH_3)_2SiCl_2(l)$$

$$SiO_2(s) + 2\ C(s) \rightarrow Si(s) + 2\ CO(g)$$

$$SiO_2(s) + 6\ HF(aq) \rightarrow SiF_6^{2-}(aq) + 2\ H^+(aq) + 2\ H_2O(l)$$

$$SiO_2(s) + 2\ NaOH(l) \rightarrow Na_2SiO_3(s) + H_2O(g)$$

$$SiO_2(s) + 3\ C(s) \rightarrow SiC(s) + 2\ CO(g)$$

$$SiO_2(s) + 2\ Na_2CO_3(l) \xrightarrow{\Delta} Na_4SiO_4(s) + 2\ CO_2(g)$$

$$2\ SiO_4^{4-}(aq) + 2\ H^+(aq) \rightarrow Si_2O_7^{6-}(aq) + H_2O(l)$$

Beyond the Basics

14.43 Lead(II) ion has a similar charge density to that of the calcium ion. Thus it will readily replace calcium ion in the bone structure.

14.45 At high temperatures, the sodium and calcium ions can leach out of the glass structure. This will result in the loss of transparency.

14.47 (a) A six-membered ring structure, Si_3O_3, with alternating silicon and oxygen atoms. Each silicon atom has two singly bonded oxygen atoms attached to it.
(b) $P_3O_9^{3-}$
(c) S_3O_9. In fact, this is the common form of sulfur trioxide, see Chapter 16.

14.49 Tin(II) chloride is the Lewis acid because it is an electron pair acceptor, while the chloride ion, the Lewis base, is an electron pair donor.

14.51 $Mg_2SiO_4(s) + 2\ "H_2CO_3(aq)" \rightarrow 2\ MgCO_3(s) + SiO_2(s) + 2\ H_2O(l)$

14.53 $A = CH_4$; $B = S$; $C = CS_2$; $D = H_2S$; $E = Cl_2$; $F = CCl_4$

$$CH_4(g) + 4\ S(s) \rightarrow CS_2(g) + 2\ H_2S(g)$$

$$CS_2(g) + 2\ Cl_2(g) \rightarrow CCl_4(g) + 2\ S(s)$$

$$CH_4(g) + 4\ Cl_2(g) \rightarrow CCl_4(g) + 4\ HCl(g)$$

14.55 To solve this problem, we need to assume that the ethyl, C_2H_5, unit remain intact. For compound Y,

$$\text{mol Sn} = 0.1240 \text{ g} \times \frac{1 \text{ mol SnO}_2}{150.5 \text{ g}} \times \frac{1 \text{ mol Sn}}{1 \text{ mol SnO}_2} = 8.239 \times 10^{-4} \text{ mol Sn}$$

We can convert to the mass of tin and subtract from the mass of compound, find the mass of ethyl units, and then determine moles. Alternatively, we can find the molar mass of the compound (assuming it contains one tin), subtract the molar mass of tin, and then determine the number of ethyl units. The latter method will be used here.

$$\text{molar mass Y} = \frac{0.1935 \text{ g}}{8.239 \times 10^{-4} \text{ mol}} = 234.9 \text{ g} \cdot \text{mol}^{-1}$$

$234.9 \text{ g} \cdot \text{mol}^{-1} = [118.7 + n(29.0)] \text{ g} \cdot \text{mol}^{-1}$

$n = 4.00$

Thus formula of Y = $Sn(C_2H_5)_4$

For compound Z, we will first work out the mass of tin(IV) oxide and silver chloride produced from 1.000 g of Z.

Mass SnO_2 = (0.1164 g)/(0.1865 g Z) = 0.6241 g

Mass AgCl = (0.1332 g)/(0.2240 g Z) = 0.5946 g

$$\text{mol Sn} = 0.6241 \text{ g} \times \frac{1 \text{ mol SnO}_2}{150.5 \text{ g}} \times \frac{1 \text{ mol Sn}}{1 \text{ mol SnO}_2} \times \frac{118.7 \text{ g}}{1 \text{ mol Sn}} = 0.4922 \text{ g Sn}$$

$$\text{mol Cl} = 0.5946 \text{ g} \times \frac{1 \text{ mol AgCl}}{143.4 \text{ g}} \times \frac{1 \text{ mol Cl}}{1 \text{ mol AgCl}} \times \frac{35.45 \text{ g}}{1 \text{ mol Cl}} = 0.1470 \text{ g Cl}$$

Mass C_2H_5 = 1.000 g – 0.4922 g – 0.1470 g = 0.3608 g C_2H_5.

$$\text{mole ratios} = \frac{0.4922 \text{ g Sn}}{118.7 \text{ g} \cdot \text{mol}^{-1}} : \frac{0.1470 \text{ g Cl}}{35.45 \text{ g} \cdot \text{mol}^{-1}} : \frac{0.3608 \text{ g C}_2\text{H}_5}{29.04 \text{ g} \cdot \text{mol}^{-1}}$$

$= 4.147 \times 10^{-3} \text{ mol Sn} : 4.147 \times 10^{-3} \text{ mol Cl} : 1.242 \times 10^{-2} \text{ mol C}_2\text{H}_5$

$= 1 \text{ Sn} : 1 \text{ Cl} : 3 \text{ C}_2\text{H}_5$

Thus formula of Z = $SnCl(C_2H_5)_3$

Equation for the reaction:

$3 \text{ Sn}(C_2H_5)_4(l) + SnCl_4(l) \rightarrow 4 \text{ SnCl}(C_2H_5)_3(l)$

This is also compatable with the information that 1.41 g of Y reacts with 0.52 g of tin(IV) chloride to give 1.93 g of Z, as the mass of Z being the sum of the masses of the reactants means there is no other product.

14.57 First we need the values for $\Delta H°$ and $\Delta S°$ of $Ca(HCO_3)_2$, which we can find by summing the values for the individual ions:

$\Delta H° = [(-543) + 2(-690)]$ kJ·mol^{-1} = -1923 kJ·mol^{-1}

$\Delta S° = [(-56) + 2(+98)]$ J·mol^{-1}·K^{-1} = $+140$ J·mol^{-1}·K^{-1}

Now we can find ΔH and ΔS for the reaction:

$\Delta H° = [(-1923) - (-1207) - (-394) - (-286)]$ kJ·mol^{-1} = -36 kJ·mol^{-1}

$\Delta S° = [(+140) - (+93) - (+214) - (+70)]$ J·mol^{-1}·K^{-1} = -237 J·mol^{-1}·K^{-1}

At 80°C:

$\Delta G° = -36$ kJ·mol^{-1} $- (353$ K$)(-0.237$ J·mol^{-1}·K$^{-1}) = +48$ kJ·mol^{-1}

A positive value indicates that the decomposition reaction will be favored.

Chapter 15

THE GROUP 15 ELEMENTS

Exercises

15.1   (a) $AsCl_3(l) + 3\ H_2O(l) \rightarrow H_3AsO_3(aq) + 3\ HCl(g)$

      (b) $3\ Mg(s) + N_2(g) \rightarrow Mg_3N_2(s)$

      (c) $NH_3(g) + 3\ Cl_2(g) \rightarrow NCl_3(l) + 3\ HCl(g)$

      (d) $CH_4(g) + H_2O(g) \rightarrow CO(g) + 3\ H_2(g)$

      (e) $N_2H_4(l) + O_2(g) \rightarrow N_2(g) + 2\ H_2O(g)$

      (f) $NH_4NO_3(aq) \rightarrow N_2O(g) + 2\ H_2O(l)$

      (g) $2\ NaOH(aq) + N_2O_3(aq) \rightarrow 2\ NaNO_2(aq) + H_2O(l)$

      (h) $2\ NaNO_3(s) \rightarrow 2\ NaNO_2(s) + O_2(g)$

      (i) $P_4O_{10}(g) + C(s) \rightarrow P_4(g) + 10\ CO(g)$

15.3   Arsenic has both metallic and nonmetallic allotropes; its oxides are amphoteric, while its chemistry mostly resembles nonmetallic phosphorus.

15.5   The first point of contrast is the difference in boiling points: the non-polar hydrocarbons have very low boiling points, while the hydrogen-bonding hydrides of nitrogen have much higher boiling points. Following from this, they have different acid-base properties, both hydrides of carbon being neutral while those of nitrogen are basic:

$$NH_3(aq) + H_2O(l) \leftrightarrow NH_4^+(aq) + OH^-(aq)$$
$$N_2H_4(aq) + H_2O(l) \leftrightarrow N_2H_5^+(aq) + OH^-(aq)$$

The most obvious difference is in their combustions, both hydrocarbons producing carbon dioxide while the nitrogen hydrides produce nitrogen gas:

$$CH_4(g) + 2\ O_2(g) \rightarrow CO_2(g) + 2\ H_2O(g)$$
$$2\ C_2H_4(g) + 6\ O_2(g) \rightarrow 4\ CO_2(g) + 4\ H_2O(g)$$
$$4\ NH_3(g) + 3\ O_2(g) \rightarrow 2\ N_2(g) + 6\ H_2O(g)$$
$$N_2H_4(l) + O_2(g) \rightarrow N_2(g) + 2\ H_2O(l)$$

15.7   (a) Nitrogen has a very strong nitrogen-nitrogen triple bond; hence any reaction that produces dinitrogen will have a strongly negative enthalpy contribution even before the other terms in the energy cycle are included. As well, being a gas, its formation will contribute a positive factor to the entropy change.

(b) Though dinitrogen is often the thermodynamically preferred product, kinetic factors (that is, comparative activation energies) can lead to other products.

15.9   Air contains about 1 percent argon. Thus, as the unreacted gases are recycled and additional air added, the argon proportion will continuously increase. One possibility would be to cool the mixture and have the argon condense out, as it has a boiling point higher than dinitrogen or dihydrogen. However, the energy requirements for this procedure would probably be prohibitive. Alternatively, the used gases could be periodically siphoned off and the dihydrogen burned as an energy source.

15.11 Though the ammonium ion is very similar to an alkali metal ion in terms of charge density (thus stabilizing large, low charge anions) and compound solubility, it has two major differences. First, a solution of the ion is acidic, not neutral, due to the following equilibrium:

$$NH_4^+(aq) + H_2O(l) \leftrightarrow H_3O^+(aq) + NH_3(aq)$$

Second, its compounds are all very thermally unstable, unlike those of the alkali metal ions, and the products are quite dissimilar. For example, ammonium nitrate decomposes on gentle heating to give dinitrogen oxide and water, while sodium nitrate upon very strong heating forms sodium nitrite and dioxygen:

$$NH_4NO_3(s) \rightarrow N_2O(g) + 2\ H_2O(g)$$
$$2\ NaNO_3(s) \rightarrow 2\ NaNO_2(s) + O_2(g)$$

15.13 The least formal charge is provided by the following two-double-bond representation

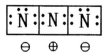

15.15 The chemical equation is

$$2\ NaN_3(s) \rightarrow 2\ Na(s) + 3\ N_2(g)$$

Mole of $NaN_3 = 7.7 \times 10^{-2}$ mol

Mole of $N_2 = 0.11(5)$ mol

Volume of gas = nRT/P

$$= (0.115\ \text{mol})(8.31\ \text{kPa·L·mol}^{-1}\text{·K}^{-1})(298K)/(102\ \text{kPa})$$
$$= 2.8\ \text{L}$$

15.17   It is ammonia that has the "anomolous" boiling point as a result of the strong hydrogen bonds between neighboring ammonia molecules.

15.19   The shapes are:

$N \equiv N - O$ (a)

(b) — a structure with two N atoms, each bonded to two O atoms

(c) — a P atom bonded to three F atoms with a lone pair

(d) — a P atom with double-bonded O, bonded to HO, OH, and H

15.21   The reaction

$$2\ NO(g) + O_2(g) \rightarrow 2\ NO_2(g)$$

is performed under high pressure because the reaction direction is favored that will result in the lesser moles of gas. A low temperature will increase the yield, as the forward process is exothermic. Both conclusions are in accordance with Le Chatelier's Principle.

15.23   White phosphorus is a very reactive, white, waxy substance that consists of P_4 molecules, shown in (a), while red phosphorus is a red powdery solid that consists of long polymer chains of linked P_4 units, shown in (b). White phosphorus burns in contact with the oxygen in air while red is air-stable. The white allotrope melts in hot water and is very soluble in non-polar solvents, while the red allotrope is high melting and is insoluble in all solvents.

(a) — a P_4 tetrahedron

(b) — linked P_4 tetrahedra forming a chain

15.25   As ammonia acts as a proton acceptor when phosphine is dissolved in it, ammonia must be the stronger base. Conversely, as phosphine acts as a proton donor, it must be a less strong base (and thus a stronger acid).

15.27 The oxidation number of nitrogen is +3, and the N–O bond order is 2:

$$:\ddot{\text{C}}\text{l}:|\text{N}|::\ddot{\text{O}}:$$

$$\quad -1 \quad +3 \quad -2$$

The chemical reaction is:

$$\tfrac{1}{2}\,N_2(g) + \tfrac{1}{2}\,O_2(g) + \tfrac{1}{2}\,Cl_2(g) \rightarrow NOCl(g)$$

Using bond energy values from the Appendix:

Bonds broken:	$\tfrac{1}{2}$(N–N)	= $\tfrac{1}{2}$(942) kJ	= 471 kJ
	$\tfrac{1}{2}$(O=O)	= $\tfrac{1}{2}$(494) kJ	= 247 kJ
	$\tfrac{1}{2}$(Cl–Cl)	= $\tfrac{1}{2}$(240) kJ	= 120 kJ
		Total	= 838 kJ
Bonds formed:	1(N–Cl)	= 1(313) kJ	= 313 kJ
	1(N–O)	= 1(N–O) kJ	= N–O kJ

Net energy change = 53 kJ = [838 – 313 – (N–O)] kJ

(N–O) = 472 kJ

The average values for single and double nitrogen-oxygen bonds are 201 kJ and 607 kJ. On this basis, the bond order is approaching that of a double bond (we have assumed that the N–Cl bond does have the typical single-bond value).

15.29 Following is the electron-dot structure and molecular shape of P_2Cl_4.

15.31 The bromine atom is probably too large for six to be accommodated around a phosphorus atom.

15.33 In hydrogen azide, formal charge arguments would give an equal weighting to the double-double and single-triple nitrogen-nitrogen bonds. As can be seen from the figure (a), the double-double nitrogen-nitrogen bond is strongly preferred.

(a)   (b)

15.35 Half reactions:

$NH_2OH(aq) + 2\ H_2O(l) \rightarrow NO_3^-(aq) + 7\ H^+(aq) + 6\ e^-$

$BrO_3^-(aq) + 6\ H^+(aq) + 6\ e^- \rightarrow Br^-(aq) + 3\ H_2O(l)$

Overall reaction

$NH_2OH(aq) + BrO_3^-(aq) \rightarrow NO_3^-(aq) + Br^-(aq) + H_3O^+(aq)$

15.37 $NOF(g) + SbF_5(l) \rightarrow NO^+(SbF_5) + SbF_6^-(SbF_5)$

15.39 $H_2S_2O_7$ and $H_6Si_2O_7$.

15.41 (a) Rapid algae growth leading to a depletion of dissolved dioxygen in lakes and rivers.

(b) Mutually beneficial relationship between two organisms.

(c) Use of a chemical compound to combat disease.

(d) Geological name of calcium hydroxide phosphate that is the bone material.

Beyond the Basics

15.43 The initial products are PH_4^+ and Cl^- (analogous to ammonia). The chloride ion then reacts with the electron-pair acceptor BCl_3 to give BCl_4^-.

$PH_3(g) + HCl(l) \rightarrow PH_4^+(HCl) + Cl^-(HCl)$

$Cl^-(HCl) + BCl_3(HCl) \rightarrow BCl_4^-(HCl)$

In the first reaction, phosphine is the Lewis base, hydrogen chloride the Lewis acid. In the second reaction, it is the chloride ion that is the base; boron trichloride is the acid.

15.45 There are two alternatives, one with the single electron in a bond, the other with it on the double-bonded oxygen. In both cases, the bond angles should all be 120°. The average bond order would be 1.33 in the first case and 1.17 in the other.

15.47 The most obvious structure would be that in which the four terminal oxygen atoms in P_4O_{10} are replaced by sulfur atoms.

15.49 The bonding between sodium and azide ions is likely to be predominantly ionic whereas that in the heavy metal azides (Fajan's rules) will be more covalent. It is the covalently bonded species, like hydrogen azide, that are more of an explosion hazard.

15.51 (a)

$$\ln K = -\frac{\Delta G^\circ}{RT} = -\frac{-16 \times 10^3 \, \text{J} \cdot \text{mol}^{-1}}{8.31 \, \text{J} \cdot \text{mol}^{-1} \cdot \text{K}^{-1} \times 298 \, \text{K}} = +6.5$$

$$K = 6 \times 10^2$$

(b) $\frac{1}{2}$ N$_2$(g) + $\frac{3}{2}$ H$_2$(g) $\leftrightarrow$ NH$_3$(g)

$\Delta H° = -46$ kJ·mol^{-1}

$\Delta S° = [(+193) - \frac{1}{2}(+192) - \frac{3}{2}(+131)]$ J·mol^{-1}·K^{-1} = -100 J·mol^{-1}·K^{-1}

$\Delta G° = \Delta H° - T\Delta S° = -46$ kJ·mol^{-1} - (775 K)(-0.100 kJ·mol^{-1}·K^{-1})

$\qquad = +32$ kJ·mol^{-1}

$\ln K = -\dfrac{+32 \times 10^3 \text{ J} \cdot \text{mol}^{-1}}{8.31 \text{ J} \cdot \text{mol}^{-1} \cdot \text{K}^{-1} \times 775 \text{ K}} = -5.0$

$K = 7 \times 10^{-3}$

(c)   The rate of the reaction is as important as the yield. By performing the reaction at a higher temperature, equilibrium is attained much more rapidly.

15.53 2 NO$_2$(g) $\leftrightarrow$ N$_2$O$_4$(g)

$\Delta H° = [(+9) - 2(+33)]$ kJ·mol^{-1} = -57 kJ·mol^{-1}

Thus the N–N bond energy must be approximately 57 kJ·mol^{-1}, assuming that the N–O bond strength remains the same.

15.55 The easiest way of solving the problem is to use the Henderson-Hasselbach equation that you should recall from your general chemistry courses:

$pH = pK_a + \log_{10} \dfrac{[\text{base}]}{[\text{acid}]}$

Let x = concentration of acid ion, then $(0.10 - x)$ = concentration of base ion.

$6.80 = 7.21 + \log \dfrac{(0.10 - x)}{x}$

$x = [\text{acid}] = 0.072$ mol·L^{-1}, [base] = 0.028 mol·L^{-1}

Mass Na$_2$HPO$_4$ = 4.0 g, mass NaH$_2$PO$_4$ = 8.6 g

15.57 The comparable bond energies (kJ·mol^{-1}) are shown below:

P–Cl   326   Cl–Cl 240

P–F   490   F–F   155

Assuming that the P–Cl bond has about the same energy in PCl$_5$ and PCl$_3$, the dissociation energy is

[240 – 2(326)] kJ·mol^{-1} = 412 kJ·mol^{-1},

a significant amount of energy, but entropy will favor decomposition at high enough temperature.

For the decomposition of PF_5, the energy change will be

$[155 - 2(490)]$ kJ·mol^{-1} = 825 kJ·mol^{-1}.

This much higher value results from fluorine bonds to other elements being stronger than those of chlorine to the same element, and the weakness of the F–F bond compared to the Cl–Cl bond.

15.59 $[NF_4]^+F^-$ — tetrafluoroammonium fluoride.

15.61   [A] Red phosphorus; [B] white phosphorus; [C] tetraphosphorus decaoxide; [D] phosphoric acid; [E] phosphorus trichloride; [F] phosphorus pentachloride; [G] phosphorous/phosphonic acid.

$4 P(s) \rightarrow P_4(s)$

$P_4(s) + 5 O_2(g) \rightarrow P_4O_{10}(s)$

$P_4O_{10}(s) + 6 H_2O(l) \rightarrow 4 H_3PO_4(aq)$

$P_4(s) + 6 Cl_2(g) \rightarrow 4 PCl_3(l)$

$PCl_3(l) + Cl_2(g) \rightarrow PCl_5(s)$

$PCl_5(s) + 4 H_2O(l) \rightarrow H_3PO_4(aq) + 5 HCl(g)$

$PCl_3(l) + 3 H_2O(l) \rightarrow H_3PO_3(aq) + 3 HCl(g)$

15.63 $Li_3N(s) + 3 H_2O(l) \rightarrow 3 LiOH(aq) + NH_3(g)$

This would be uneconomical, as one would need to produce lithium metal first, a highly energy-intensive electrolytic process, in order to synthesize the lithium nitride in the first instance:

$6 Li(s) + N_2(g) \rightarrow 2 Li_3N(s)$

15.65 $HONH_2$ (or NH_2OH, hydroxylamine); H_2NNO_2; $(NH_2)_2CO$ (urea).

15.67 $2\ NCl_3(g) \rightarrow N_2(g) + 3\ Cl_2(g)$

Using bond energy values from the Appendix:

Bonds broken:	6(N–Cl)	=6(192 kJ)	= 1152 kJ
		Total	= 1152 kJ
Bonds formed:	(N≡N)		= 942 kJ
	3(Cl–Cl)	= 3(240) kJ	= 720 kJ
		Total	= 1662 kJ

Net energy change $= +1152\ kJ - 1662\ kJ = -510\ kJ$

The reaction is highly exothermic due primarily to the strength of the nitrogen-nitrogen triple bond.

15.69   Only two hydrogens are replaced as the structure only contains two hydroxyl groups. The hydrogen bonded to the phosphorus is not labile and cannot be replaced.

$(HO)_2HPO_2(D_2O) + 2\ D_2O(l) \rightarrow (DO)_2HPO_2(D_2O) + 2\ HDO(D_2O)$

15.71 NO_2^+ and CNO^- are isoelectronic with the azide ion.

15.73   A very large low-charge anion – such as the hexafluoroosmate ion, $[OsF_6]^-$ ion – might stabilize the pentanitrogen cation.

15.75   (a) Silver(I) or lead(II) or mercury(I).

   (b) $N_3^-(aq) + H_2O(l) \rightarrow HN_3(aq) + OH^-(aq)$

   (c) The azide ion will decompose on heating to give nitrogen gas.

15.77   We can write half-reactions as if the reaction took place in solution, provided that when the final balanced equation is derived, the hydrogen ions and electrons "cancel out."

$(NH_4)[N(NO_2)_2](s) + 4\ H^+(aq) + 4\ e^- \rightarrow 4\ H_2O(g) + 2\ N_2(g)$

$2\ Al(s) + 3\ H_2O(g) \rightarrow Al_2O_3(s) + 6\ H^+(aq) + 6\ e^-$

Balanced equation:

$3\ (NH_4)[N(NO_2)_2](s) + 4\ Al(s) \rightarrow 2\ Al_2O_3(s) + 6\ H_2O(g) + 6\ N_2(g)$

15.79 Assume arsenic is present as the arsenate ion, $AsO_4^{3-}(aq)$

$$Zn(s) + H_2SO_4(aq) \rightarrow Zn^{2+}(aq) + SO_4^{2-}(aq) + H_2(g)$$

$$AsO_4^{3-}(aq) + 4\ H_2(g) \rightarrow AsH_3(g) + 3\ OH^-(aq) + H_2O(l)$$

$$2\ AsH_3(g) \xrightarrow{\ \Delta\ } 2\ As(s) + 3\ H_2(g)$$

Chapter 16

THE GROUP 16 ELEMENTS

Exercises

16.1 (a) $2 Fe(s) + 3 O_2(g) \rightarrow 2 Fe_2O_3(s)$

(b) $BaS(s) + 4 O_3(s) \rightarrow BaSO_4(s) + 4 O_2(g)$

(c) $BaO_2(s) + 2 H_2O(l) \rightarrow Ba(OH)_2(aq) + H_2O_2(aq)$

(d) $2 KOH(aq) + CO_2(g) \rightarrow K_2CO_3(aq) + H_2O(l)$
    $K_2CO_3(aq) + CO_2(g) + H_2O(l) \rightarrow 2 KHCO_3(aq)$

(e) $Na_2S(aq) + H_2SO_4(aq) \rightarrow Na_2SO_4(aq) + H_2S(g)$

(f) $Na_2SO_3(aq) + H_2SO_4(aq) \rightarrow Na_2SO_4(aq) + SO_2(g) + H_2O(l)$

(g) $8 Na_2SO_3(aq) + S_8(s) \rightarrow 8 Na_2S_2O_3(aq)$

16.3 Polonium is the only element of the group to have an electrical resistivity low enough to be considered metallic.

16.5 (a) Finely divided metals that are spontaneously flammable in air.
(b) Different crystal forms of an element in which identical molecular units are packed differently.
(c) Unusual type of equilibria found with hemoglobin in which addition of one oxygen molecule increases the ease of addition of subsequent oxygen molecules.

16.7 Photosynthesis has resulted in the conversion of most of the carbon dioxide that was in the Earth's atmosphere to dioxygen.

16.9 The bond order should be that in ozone itself—about 1½. Using the molecular orbital approach, the electron would be removed from one of the nonbonding orbitals, so there would be no effect on bond order. From simple bonding models, we obtain the same result.

Also, by comparison with the isoelectronic nitrogen dioxide molecule, we would argue that the electron would be lost from the central oxygen's lone pair, opening out the bond angle to greater than 120°.

16.11 We would expect a larger angle for dibromine oxide compared to dichlorine oxide, both from the Bent rule and from consideration of steric crowding by the two large bromine atoms.

16.13 See the diagram. The oxidation number of +1 for oxygen is extremely rare and is a result of each atom being sandwiched between a more electronegative fluorine atom on one side and an identical atom of oxygen on the other.

$$:\!\ddot{F}\!:\!\ddot{O}\!:\!\ddot{O}\!:\!\ddot{F}\!:$$
$$\;\;-1\;\;+1\;\;+1\;\;-1$$

16.15 Among the Group 16 elements, it is only sulfur that readily catenates. With the high stability of oxygen multiple bonds, it is extremely rare to find chains of oxygen atoms. It is probable that the high electronegativity of the end group "pulls" electron density away from the oxygen chain and in some way stabilizes it.

16.17

(a) (b) (c) (d)

16.19 The structure is probably based on the S_8 ring. As the NH combination has the same number of outer electrons as sulfur, it could simply replace alternate sulfur atoms around the ring and retain the same electron configuration.

16.21 The terminal oxygen atoms have "normal" oxidation states of −2, while sulfur has the common value of +6. The linked oxygen atoms are peroxide-type with oxidation states of −1.

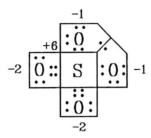

16.23 Whitewash was effective because the calcium hydroxide reacted with carbon dioxide from the air to form opaque calcium carbonate:

$$Ca(OH)_2(s) + CO_2(g) \rightarrow CaCO_3(s) + H_2O(l)$$

16.25 As the yellow solid sulfur is heated, it first melts to give a low-viscosity, straw-colored liquid. This consists of S_8 molecules. As the temperature is increased, the liquid darkens and becomes very viscous. Over this range, the S_8 rings are breaking open to form long polymer chains. At higher temperatures still, the viscosity decreases as the average chain length decreases. When the liquid boils gaseous green, S_8 molecules are produced, and at even higher temperatures, the gas turns violet as the rings break into S_2 molecules analogous to O_2.

16.27 The closeness of the bond angle in H_2Te to 90° suggests that the central tellurium atom is using pure *p* orbitals in its bonding.

16.29 Sulfuric acid can act as an acid (usually when dilute); as a dehydrating agent (when concentrated), as an oxidizing agent (when hot and concentrated), as a sulfonating agent (when concentrated), and as a base with stronger acids.

16.31 The most likely formal charge arrangements for sulfur trioxide and the sulfite ion are shown in (a) and (b), respectively. In the former, the average bond is $1^2/_3$, while that in the latter is $1^1/_3$. Thus sulfur trioxide will have the stronger bonds (this has been established—the S–O bond length is sulfur trioxide is 142 pm while that in the sulfite ion is 151 pm).

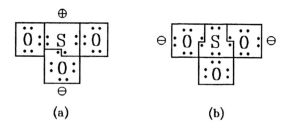

(a)   (b)

16.33

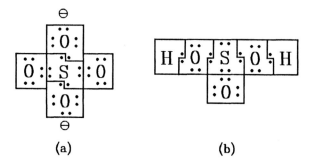

(a)   (b)

16.35 (a) A black precipitate of lead(II) sulfide is produced from a colorless lead(II) acetate solution:

$$H_2S(g) + Pb(CH_3COO)_2(aq) \rightarrow PbS(s) + 2\ CH_3COOH(aq)$$

(b) A white precipitate is produced when a solution of barium ion is added to sulfate ion:

$$Ba^{2+}(aq) + SO_4^{2-}(aq) \rightarrow BaSO_4(s)$$

16.37 Though the oxidation of sulfur dioxide to sulfur trioxide is thermodynamically spontaneous, it is kinetically very slow—or in other words, there is a very high activation energy barrier to reaction.

16.39 The large tetramethylammonium cation will stabilize the large, low-charge ozonide ion.

16.41   The NS_2^+ ion is isoelectronic and isostructural with carbon disulfide, CS_2.

$$\left[\ddot{S} :: N :: \ddot{S}\right]^+$$

16.43   We require small quantities of selenium for a healthy existence, but in larger quantities in our diet, it is highly toxic.

Beyond the Basics

16.45   See the diagrams. The value of −668 kJ is much less than the −1209 kJ for sulfur hexafluoride. This difference is accounted for by the chlorine-chlorine bond being stronger than the fluorine-fluorine bond and the sulfur-chlorine bond being weaker than the sulfur-fluorine bond.

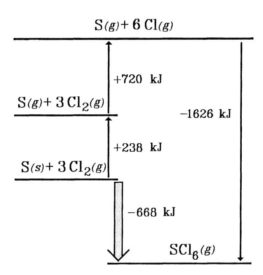

16.47   The ammonium salt will be less basic than the calcium salt as the ammonium ion is the conjugate base of a weak acid, so it reacts with water to produce a hydronium ion, partially neutralizing the hydroxide ion produced from the reaction of the glycollate ion with water to produce its conjugate acid.

$$NH_4^+(aq) + H_2O(l) \leftrightarrow H_3O^+(aq) + NH_3(aq)$$

16.49 Concentration in mol·cm^{-3} = $\dfrac{5\times10^5 \text{ molecules}}{\text{cm}^3} \times \dfrac{1\,\text{mol}}{6.02\times10^{23}\text{ molecules}}$

= 8 × 10^{-19} mol·cm^{-3} = 8 × 10^{-16} mol·L^{-1}

At SATP, 1 L of gas contains n = PV/RT mol of gas.

$$n = \dfrac{100\,\text{kPa}\times1\,\text{L}}{8.31\,\text{kPa}\cdot\text{L}\cdot\text{mol}^{-1}\cdot\text{K}^{-1}\times298\,\text{K}} = 0.04\,\text{mol}$$

Concentration in ppb = $\dfrac{8\times10^{-16}\,\text{mol}}{0.04\,\text{mol}}\times10^9 = 2\times10^{-5}$ ppb

16.51 (a) Length of side = 2(74 pm) + 2(126 pm) = 400 pm

(b) Length of diagonal from mid-edge through center = 2(126) + 2(114 pm) = 480 pm

Thus length of side = 1/√2(480 pm) = 339 pm.

16.53 The S=O double bond is preferred according to formal charge principles (see the diagram). The molecule will be trigonal planar.

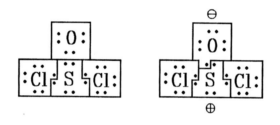

16.55 S(s) + O$_2$(g) → SO$_2$(g) and CaCO$_3$(s) + SO$_2$(g) → CaSO$_3$(s) + CO$_2$(g)

mol S = 1000 tonne coal × $\dfrac{3 \text{ parts S}}{100 \text{ parts coal}} \times \dfrac{10^6\,\text{g}}{1 \text{ tonne}} \times \dfrac{1 \text{ mol S}}{32.1\,\text{g}} = 9.3\times10^5$ mol S

mass CaCO$_3$ = 9.3×10^5 mol S × $\dfrac{1 \text{ mol SO}_2}{1 \text{ mol S}} \times \dfrac{1 \text{ mol CaCO}_3}{1 \text{ mol SO}_2} \times \dfrac{100.1\,\text{g}}{1 \text{ mol CaCO}_3} \times \dfrac{1 \text{ tonne}}{10^6\,\text{g}}$

= 94 tonne

16.57 Apparent oxidation number S

$0 = 2[H] + [S] + 5[O]$

$0 = 2[+1] + [S] + 5[-2]$

[S] = +8, an impossible value as the oxidation number of sulfur cannot exceed 6.

$H_2SO_5 + H_2O \rightarrow H_2SO_4 + H_2O_2$ is the only equation that fits the hint.

Thus a likely structure is that of sulfuric acid, $(HO)_2SO_2$ but with a peroxy- link instead of a hydroxy- link.

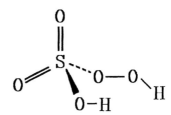

16.59 It is most probable that sulfite is oxidized to sulfate while peroxodisulfate is reduced to sulfate.

$SO_3^-(aq) + H_2O(l) \rightarrow SO_4^{2-}(aq) + 2 H^+(aq) + 2 e^-$

$S_2O_8^{2-}(aq) + 2 e^- \rightarrow 2 SO_4^{2-}(aq)$

$SO_3^{2-}(aq) + S_2O_8^{2-}(aq) + H_2O(l) \rightarrow 3 SO_4^{2-}(aq) + 2 H^+(aq)$

16.61 $E = E° - \dfrac{RT}{2F} \ln \dfrac{1}{[H^+]^4}$

$E = +0.17\ V - \dfrac{(8.31\ C \cdot V \cdot mol^{-1} \cdot K^{-1})(298\ K)}{(2)(9.65 \times 10^4\ C \cdot mol^{-1})} \ln\left(\dfrac{1}{(1.0 \times 10^{-14})^4} \right)$

$E = +0.17\ V - 1.65\ V = -1.48\ V$

16.63 [A] Sulfur dioxide; [B] potassium hydroxide; [C] potassium sulfite; [D] sulfur; [E] thiosulfate ion; [F] tetrathionate ion; [G] thiosulfuric acid.

$SO_2(g) + 2 KOH(aq) \rightarrow K_2SO_3(aq)$

$K^+(aq) + [B(C_6H_5)_4]^-(aq) \rightarrow K[B(C_6H_5)_4](s)$

$K_2SO_3(aq) + S(s) \rightarrow K_2S_2O_3(aq)$

$2 S_2O_3^{2-}(aq) + I_2(aq) \rightarrow S_4O_6^{2-}(aq) + 2 I^-(aq)$

$S_2O_3^{2-}(aq) + 2 H^+(aq) \rightarrow H_2S_2O_3(aq)$

$H_2S_2O_3(aq) \rightarrow H_2O(l) + S(s) + SO_2(g)$

16.65 The triple bond structure is more likely as there are no formal charges in the electron arrangement.

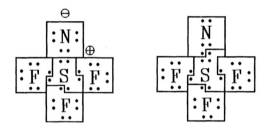

16.67 Rubidium or cesium. A large low-charge cation is necessary to stabilize the low-charge anion.

16.69 In SF_2, the sulfur has a formal oxidation state of +2. In F_3SSF, the sulfur surrounded by three fluorines will have the higher oxidation state of +3 (note the lone pair on this sulfur atom), while the sulfur attached to only one fluorine will have an oxidation state of +1.

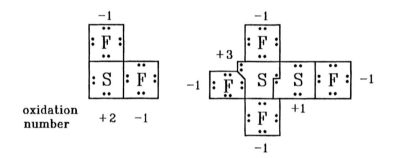

Chapter 17

THE GROUP 17 ELEMENTS: THE HALOGENS

Exercises

17.1   (a) $UO_2(s) + 4\ HF(g) \rightarrow UF_4(s) + 2\ H_2O(l)$

       (b) $CaF_2(s) + H_2SO_4(l) \rightarrow 2\ HF(g) + CaSO_4(s)$

       (c) $SCl_4(l) + 2\ H_2O(l) \rightarrow SO_2(g) + 4\ HCl(g)$

       (d) $3\ Cl_2(aq) + 6\ NaOH(aq) \rightarrow NaClO_3(aq) + 5\ NaCl(s) + 3\ H_2O(l)$

       (e) $I_2(s) + 5\ F_2(g) \rightarrow 2\ IF_5(s)$

       (f) $BrCl_3(l) + 2\ H_2O(l) \rightarrow 3\ HCl(aq) + HBrO_2(aq)$

17.3   Fluorine has a very weak fluorine-fluorine bond; it is usually limited to one or two covalent bonds; its compounds with metals are often ionic when those of the comparable chlorides are covalent; it has an extremely high electronegativity and forms the strongest hydrogen bonds known; it tends to stabilize high oxidation states; the solubility of its metal compounds for a particular metal is often quite different than those of the other halides.

17.5   The reaction with nonmetals is strongly enthalpy-driven as weak fluorine-fluorine bonds are broken and strong bonds are formed between fluorine and other nonmetals.

17.7   The chemical equation is

        $I_2(s) + 7\ F_2(g) \rightarrow 2\ IF_7(s)$

     Thus there is a decrease of seven moles of gas in this spontaneous reaction. Hence enthalpy decrease must be the driving force of the reaction.

17.9   The half-equation is:

        $\frac{1}{2}\ Cl_2(aq) + e^- \rightarrow Cl^-(aq)$

     As hydrogen ion does not appear in the half-equation, the reduction potential will not be pH sensitive.

17.11 In the ionization of a hydrohalic acid

$$HX(aq) + H_2O(l) \rightarrow H_3O^+(aq) + X^-(aq)$$

the H–X bond must be broken. The H–F bond is particularly strong, and as a result, the equilibrium will tend to lie to the left and hydrofluoric acid will behave as a weak acid.

17.13 Chemical equation:

$$CaF_2(s) + H_2SO_4(l) \rightarrow CaSO_4(s) + 2\ HF(g)$$

Mass of hydrogen fluoride = 1.2×10^{12} g

Moles of hydrogen fluoride = 6.0×10^{10} mol

Moles of calcium sulfate = 3.0×10^{10} mol

Mass of calcium sulfate = 4.1×10^{12} g = 4.1×10^6 tonne

17.15 Oxygen has the unusual oxidation number of zero in this compound.

17.17 (a) To form the higher oxidation state of a metal, dichlorine should be used:

$$2\ Cr(s) + 3\ Cl_2(g) \rightarrow 2\ CrCl_3(s)$$

(b) To form the lower oxidation state of a metal, iodine monochloride should be used:

$$Cr(s) + 2\ ICl(l) \rightarrow CrCl_2(s) + I_2(s)$$

17.19 Iron(III) iodide will not be stable as iodide ion is a reducing agent, hence it will reduce iron(III) to iron(II):

$$2\ I^- \rightarrow I_2 + 2\ e^-$$
$$Fe^{3+} + e^- \rightarrow Fe^{2+}$$

17.21 The chemical equation is

$$6\ NH_4ClO_4(s) + 8\ Al(s) \rightarrow 4\ Al_2O_3(s) + 3\ N_2(g) + 3\ Cl_2(g) + 12\ H_2O(g)$$

Thus $\Delta H = 12[\Delta H_f(H_2O(g))] + 4[\Delta H_f(Al_2O_3(s)) - 6[\Delta H_f(NH_4ClO_4(s))]$
(as the other species are elements in their standard phases)

$\Delta H \quad = 12[-242\ kJ] + 4[-1676\ kJ] - 6[+295\ kJ]$

$\quad\quad = -7838\ kJ$

It would also be a good propellant because it produces a large number of small gas molecules; thus it would have a high thrust per unit mass.

17.23   The half-equations for the first reaction
$$H_2S(g) + 2\ H_2O(l) \rightarrow SO_2(g) + 6\ H^+(aq) + 6\ e^-$$
$$I_2O_5(s) + 10\ H^+(aq) + 10\ e^- \rightarrow I_2(s) + 5\ H_2O(l)$$
Give an overall equation of
$$10\ H_2S(g) + 6\ I_2O_5(s) \rightarrow 10\ SO_2(g) + 6\ I_2(s) + 10\ H_2O(l)$$
For the second reaction:
$$I_2(s) + 2\ S_2O_3^{2-}(aq) \rightarrow 2\ I^-(aq) + S_4O_6^{2-}(aq)$$

17.25   Three arguments can be used: first, that the sulfur atom is too small to accomodate six iodine atoms around it; second, that iodine is reducing, thus such a high oxidation state cannot be stabilized; and third, the sulfur-iodine bond energy is not sufficient to provide an exothermic balance to the decrease in entropy that would result from consuming six moles of gas per mole of compound formed.

17.27   The central chlorine atom has a +7 oxidation number and the end chlorine has a +1 oxidation number. Both oxidation numbers are common for chlorine.

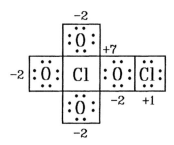

17.29   As iodine can almost be classed as a semimetal (for example, it is known in the +1 oxidation state), we would predict astatine to start to show some metallic properties; thus the diatomic element might be a significant electrical conductor. Like the other halogens, astatine should have a common oxidation state of –1 and form an insoluble compound with silver ion:
$$Ag^+(aq) + At^-(aq) \rightarrow AgAt(s)$$
The compound should be insoluble in concentrated ammonia solution. All of the other halogens should displace its anion. For example, iodine should react as follows:
$$I_2(aq) + 2\ At^-(aq) \rightarrow 2\ I^-(aq) + At_2(s)$$

Astatine should form interhalogen compounds, such as AtF and AtI, in which astatine has a positive polarity. In fact, being so near the metal/nonmetal border, astatine will probably have a significant cation chemistry, forming perhaps, At^+ and At^{3+}.

17.31 Structure (c) with the charge on the sulfur atom must be the major contributor with, possibly, some small contribution from structure (a). In view of the high formal charge, contributions from structure (b) can be ignored.

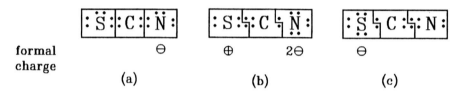

formal charge

(a)   (b)   (c)

17.33 Fluorine

$Cl_2(g) + 3 F_2(g) \rightarrow 2 ClF_3(g)$

$S(s) + 3 F_2(g) \rightarrow SF_6(g)$

$BrO_3^-(aq) + F_2(g) + 2 OH^-(aq) \rightarrow BrO_4^-(aq) + 2 F^-(aq) + H_2O(l)$

$2 Fe(s) + 3 F_2(g) \rightarrow 2 FeF_3(s)$

$H_2(g) + F_2(g) \rightarrow 2 HF(g)$

$2 F^-(KH_2F_3) \rightarrow F_2(g) + 2 e^-$

$HF(aq) + OH^-(aq) \rightarrow H_2O(l) + F^-(aq)$

$HF(aq) + F^-(aq) \rightarrow HF_2^-(aq)$

$6 HF(aq) + SiO_2(s) \rightarrow SiF_6^{2-}(aq) + 2 H^+(aq) + 2 H_2O(l)$

$4 HF(g) + UO_2(s) \rightarrow UF_4(s) + 2 H_2O(g)$

$UF_4(s) + F_2(g) \rightarrow UF_6(g)$

Chlorine

$P_4(s) + 10 Cl_2(g) \rightarrow 4 PCl_5(s)$

$2 Fe(s) + 3 Cl_2(g) \rightarrow 2 FeCl_3(s)$

$3 Cl_2(g) + NH_3(g) \rightarrow NCl_3(l) + 3 HCl(g)$

$Cl_2(aq) + 2 OH^-(aq) \rightarrow Cl^-(aq) + ClO^-(aq) + H_2O(l)$

$ClO^-(aq) + H^+(aq) \rightarrow HClO(aq)$

$2 ClO^-(aq) + Ca^{2+}(aq) \rightarrow Ca(ClO)_2(s)$

$Cl_2(g) + H_2(g) \rightarrow 2 HCl(g)$

$2 HCl(g) + Fe(s) \rightarrow FeCl_2(s) + H_2(g)$

$3 Cl_2(aq) + 6 OH^-(aq) \rightarrow ClO_3^-(aq) + 5 Cl^-(aq) + 3 H_2O(l)$

$$ClO_3^-(aq) + H_2O(l) \rightarrow ClO_4^-(aq) + 2\ H^+(aq) + 2\ e^-$$
$$2\ ClO_3^-(aq) + 4\ H^+(aq) + 2\ Cl^-(aq) \rightarrow 2\ ClO_2(aq) + Cl_2(g) + 2\ H_2O(l)$$

Iodine

$$I_2(s) + Cl_2(g) \rightarrow 2\ ICl(s)$$
$$I_2(s) + 2\ S_2O_3^{2-}(aq) \rightarrow 2\ I^-(aq) + S_4O_6^{2-}(aq)$$
$$2\ I^-(aq) + Cl_2(g) \rightarrow I_2(aq) + 2\ Cl^-(aq)$$
$$I^-(aq) + I_2(aq) \rightarrow I_3^-(aq)$$

17.35   Chlorine oxidation state = +1, oxygen = –1.

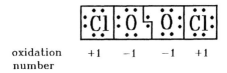

oxidation   +1   –1   –1   +1
number

17.37   The large low-charge iodide anion will stabilize the large low-charge cation.

17.39   Bromine monofluoride, BrF, would be an analog of dichlorine, Cl_2. Note that here we are using the term "combo" element as a combination of elements of the same group but of period $(n-1)$ and $(n+1)$ as analogs of the element in period n.

Beyond the Basics

17.41   The ammonium hydrogen fluoride may be decomposing and dissolving in the hydrogen fluoride produced.
$$(NH_4)^+(HF_2)^-(s) \rightarrow NH_4^+\ (HF) + F^-(HF)$$

17.43 The bond order will be (3 − 2) = 1.

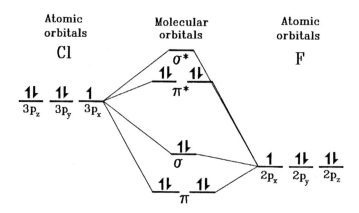

17.45 Dichlorine heptaoxide. It is the oxide in the higher oxidation state (with more oxygen atoms) that will be acidic.

17.47 Each chlorine atom will be approximately tetrahedrally coordinated, one chlorine with one lone pair and the other chlorine with two. Thus the bond angles will be approximately 109½°. Though the electron-dot structure depicts singly bonded oxygens, there is certainly multiple bond character.

17.49 The perchlorate ion is a strong oxidizing agent, but it needs to be mixed with an oxidizable compound or element in order to have explosive properties. In ammonium perchlorate, the easily oxidizable ammonium ion is an integral part of the compound. Thus no additional component is required to cause a vigorous redox reaction.

$2 NH_4ClO_4(s) \rightarrow N_2(g) + Cl_2(g) + 2 O_2(g) + 4 H_2O(g)$

Nitrogen is oxidized from −3 to 0, chlorine reduced from +7 to 0, oxygen is oxidized from −2 to 0.

17.51 The structures follow. Single Cl–O bonds would be equally acceptable.

17.53 $Tl^+(I_3)^-$. Iodide is a reducing agent and it would be expected to reduce thallium(III) to thallium(I).

17.55   (a)   The azide, N_3^- ion, acts as a pseudo-halide ion. Thus it can form a pseudo-interhalide ion by substituting for two of the iodine atoms in I_3^-, thus $[I(N_3)_2]^-$.

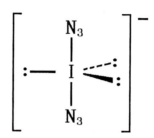

(b)   The azide must be of higher electronegativity than iodide as it is the less electronegative element that is the center of a cluster.

(c)   The iodide ion has seven electrons plus one from the negative charge. It will "share" one with each of the azide units, giving a total of ten electrons, or five electron pairs. Hence a trigonal bipyramid electron-pair arrangement. As the three lone pairs will occupy equatorial positions, the molecular shape will be linear.

(d)   The ion would be stabilized in the solid phase by a larege cation such as rubidium or cesium.

17.57   (a) $ClF_3(l) + BF_3(g) \leftrightarrow ClF_2^+(ClF_3) + BF_4^-(ClF_3)$

(b) $ClF_3(l) + KF(s) \leftrightarrow K^+(ClF_3) + ClF_4^-(ClF_3)$

(c)   In (a), the B–F bond is much stronger (613 kJ·mol^{-1}) than the Cl–F bond (249 kJ·mol^{-1}), thus the boron can abstract a fluorine from the chlorine. In (b), the Cl–F bond strength must be greater than the energy needed to extract a fluoride ion from the potassium fluoride lattice.

Chapter 18

THE GROUP 18 ELEMENTS: THE NOBLE GASES

Exercises

18.1 (a) $Xe(g) + 2 F_2(g) \rightarrow XeF_4(s)$

(b) $XeF_4(s) + 2 PF_3(g) \rightarrow 2 PF_5(g) + Xe(g)$

18.3 Descending the group, the melting and boiling points increase, as does the density.

18.5 Helium cannot be solidified at any temperature under normal pressure; when cooled close to absolute zero, liquid helium (helium II) becomes an incredible thermal conductor and its viscosity drops to close to zero.

18.7 Assuming the molecular orbitals formed from $5p$ atomic orbitals are similar to those formed from $2p$ atomic orbitals, we can construct the following diagram and deduce that the bond order must be ½.

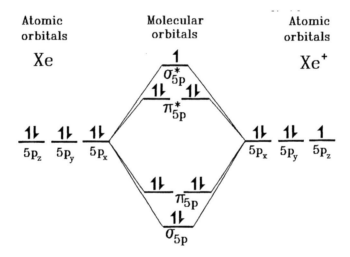

18.9 The thermodynamic factors are the weakness of the fluorine-fluorine bond that has to be broken in the energy cycle and the comparative strength of the xenon-fluorine bond that is formed.

18.11

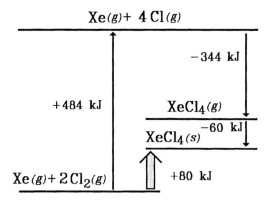

18.13 The double-bond structure has zero formal charge while the single-bond representation has a negative charge on the oxygen and a positive charge on the xenon. Thus the double-bonded structure probably makes a major contribution to the bonding.

18.15 Using the calculation method:

(a) $[N_{ox}(Xe)] + 3(-1) = +1$
$[N_{ox}(Xe)] = +4$

(b) $[N_{ox}(Xe)] + 5(-1) = +1$
$[N_{ox}(Xe)] = +6$

(c) $[N_{ox}(Xe)] + 6(-2) = -4$
$[N_{ox}(Xe)] = +8$

18.17 Rubidium or cesium as a large, low charge cation is needed to stabilize a large low charge anion.

18.19 Half reactions:

$Au \rightarrow Au^{5+} + 5\ e^-$

$KrF_2 + 2\ e^- \rightarrow Kr + 2\ F^-$

Redox reaction:

$2\ Au + 5\ KrF_2 \rightarrow 2\ AuF_5 + 5\ Kr$

Plus two extra KrF_2 to result in salt formation:

$2\ Au + 7\ KrF_2 \rightarrow 2\ (KrF)^+(AuF_6^-) + 5\ Kr$

18.21 $Xe(g) + F_2(g) \rightarrow XeF_2(s)$

$2\,XeF_2(s) + 2\,H_2O(l) \rightarrow 2\,Xe(g) + O_2(g) + 4\,HF(l)$

$Xe(g) + 2\,F_2(g) \rightarrow XeF_4(s)$

$Xe(g) + 3\,F_2(g) \rightarrow XeF_6(s)$

$XeF_6(s) + H_2O(l) \rightarrow XeOF_4(l) + 2\,HF(l)$

$XeOF_4(l) + 2\,H_2O(l) \rightarrow XeO_3(s) + 4\,HF(l)$

$XeO_3(s) + OH^-(aq) \rightarrow HXeO_4^-(aq)$

$2\,HXeO_4^-(aq) + 2\,OH^-(aq) \rightarrow XeO_6^{4-}(aq) + Xe(g) + O_2(g) + H_2O(l)$

$XeO_6^{4-}(aq) + 2\,Ba^{2+}(aq) \rightarrow Ba_2XeO_6(s)$

$Ba_2XeO_6(s) + 2\,H_2SO_4(aq) \rightarrow 2\,BaSO_4(s) + XeO_4(g) + 2\,H_2O(l)$

Beyond the Basics

18.23 $XeF_2(SbF_5) + SbF_5(l) \leftrightarrow XeF^+(SbF_5) + SbF_6^-(SbF_5)$

18.25 $Ar(g) + F_2(g) \rightarrow ArF_2(g)$

Even assuming argon difluoride would be a gas, there is a decrease in entropy in the synthesis. Thus the enthalpy change has to be less than zero (exothermic). Assuming zero is the limiting value, we can construct an energy cycle. Thus the Ar–F bond energy can be no more than 77.5 kJ·mol^{-1}.

Chapter 19

INTRODUCTION TO TRANSITION METAL COMPLEXES

Exercises

19.1 (a) Element belonging to the d-block, though usually groups 3 and 12 are excluded.

(b) Molecules or ions covalently bonded to a central metal ion.

(c) Energy separation between different members of the metal's d-orbital set when the metal ion is surrounded by a set of ligands.

19.3 The cyanide ligand stabilizes low oxidation states (in a similar manner to carbon monoxide) and also stabilizes normal oxidation states (as a pseudohalide ion).

19.5 $[Pt(NH_3)_4]^{2+}[PtCl_4]^{2-}$

19.7

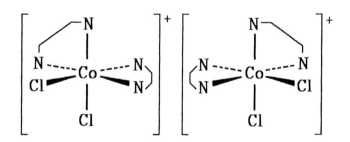

19.9 (a) Ammonium pentachlorocuprate(II); (b) pentaammineaquacobalt(III) bromide; (c) potassium tetracarbonylchromate(-III); (d) potassium hexafluoronickelate(IV); (e) tetraamminecopper(II) perchlorate.

19.11 (a) $[Mn(OH_2)_6](NO_3)_2$.
(b) $Pd[PdF_6]$.
(c) $[CrCl_2(OH_2)_4]Cl \cdot 2H_2O$.
(d) $K_3[Mo(CN)_8]$.

19.13 (a) The d^6 configuration in an octahedral field:

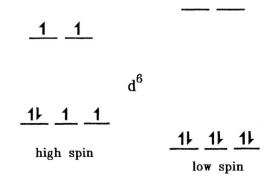

(b) The d^6 configuration in a tetrahedral field:

19.15 The largest value of Δ is for the cobalt(III) complex, the others being cobalt(II) as the splitting increases with increase in oxidation state. The smallest value is for cobalt(II) in a tetrahedral environment compared to the middle cobalt(II) in an octahedral environment as Δ_{tet} is only about four-ninths the value of Δ_{oct}.

19.17 (a) $[ReF_6]^{2-}$ as the heavier metal will have the greater crystal field splitting.
(b) $[Fe(CN)_6]^{3-}$ as the higher charge Fe(III) will have the greater crystal field splitting.

19.19

Configuration	CFSE
d^0	$-0.0\ \Delta_{tet}$
d^1	$-0.6\ \Delta_{tet}$
d^2	$-1.2\ \Delta_{tet}$
d^3	$-0.8\ \Delta_{tet}$
d^4	$-0.4\ \Delta_{tet}$
d^5	$-0.0\ \Delta_{tet}$
d^6	$-0.6\ \Delta_{tet}$
d^7	$-1.2\ \Delta_{tet}$
d^8	$-0.8\ \Delta_{tet}$
d^9	$-0.4\ \Delta_{tet}$
d^{10}	$-0.0\ \Delta_{tet}$

19.21 The optimum situation energetically is for the ion with the greater CFSE to occupy the octahedral sites. Thus the mixed metal oxide $NiCr_2O_4$ will adopt the normal spinel structure, $(Ni^{2+})_t(Cr^{3+})_oO_4$ because the Cr^{3+} ion, having the higher oxidation state, will have a greater CFSE than that of the Ni^{2+} ion.

19.23 Balanced chemical equation:
$$[Ni(OH_2)_6]^{2+}(aq) + 2\ det(aq) \rightarrow [Ni(det)_2]^{2+}(aq) + 6\ H_2O(l)$$
The formation of this product will be favored as a result of the chelate effect— the increase in entropy from the increase in moles.

Beyond the Basics

19.25 The ligand tricyclohexylphosphine is probably too large for two to fit around an iron(III) in addition to the three chloro-ligands.

19.27 (a) M^{2+} should disproportionate as the sum of the potentials $(0.00 + 0.20)$ V is positive. The equation would be
$$3\ M^{2+}(aq) \rightarrow M(s) + 2\ M^{3+}(aq)$$

(b) $2 M^{2+}(aq) + 2 H^+(aq) + 2 e^- \rightarrow 2 M^{3+}(aq) + H_2(g)$

To find the limit of spontaneity, we can set E = 0.

As only the hydrogen ion concentration varies we can write a simplified Nernst equation:

$$E = 0 = E^\circ - \frac{RT}{2F} \ln\left(\frac{1}{[H^+]^2}\right) = +0.20 \text{ V} - \frac{8.31 \text{ V}\cdot\text{C}\cdot\text{mol}^{-1}\cdot\text{K}^{-1} \times 298 \text{ K}}{2 \times (9.65 \times 10^4 \text{ C}\cdot\text{mol}^{-1})} \ln\left(\frac{1}{[H^+]^2}\right)$$

$$[H^+] = 4.1 \times 10^{-4}$$

$$pH = 3.38$$

19.29 For zinc, with its filled d^{10} orbitals, there is no crystal field stabilization energy; thus geometry is primarily determined by geometry. For nickel, a square-planar geometry will enable some degree of π bonding to occur between the part-empty d orbitals of the nickel and the filled d orbitals of the selenium.

19.31 (A) $[Cr(OH_2)_6]^{3+}\cdot3Cl^-$, hexaaquachromium(III) chloride,

(B) $[Cr(OH_2)_5Cl]^{2+}\cdot2Cl^-$, pentaaquachlorochromium(III) chloride,

(C) $[Cr(OH_2)_4Cl_2]^+\cdot Cl^-$, tetraaquadichlorochromium(III) chloride.

19.33 Fluoride is a weaker field ligand than chloride. Thus the crystal field splitting for the fluoro- compound will be less than that of the chloro- compound. To give a yellow-orange color, the chloro- compound must be absorbing in the blue (higher energy) portion of the spectrum. To give a blue color, the fluoro- compound must be absorbing in the red (lower energy) portion of the spectrum.

Chapter 20

PROPERTIES OF THE TRANSITION METALS

Exercises

20.1 (a) $TiCl_4(l) + O_2(g) \rightarrow TiO_2(s) + 2 Cl_2(g)$

(b) $Na_2Cr_2O_7(s) + S(l) \rightarrow Cr_2O_3(s) + Na_2SO_4(s)$

(c) $Cu(OH)_2(s) \rightarrow CuO(s) + H_2O(l)$

(d) $2 [Ag(CN)_2]^-(aq) + Zn(s) \rightarrow 2 Ag(s) + [Zn(CN)_4]^{2-}(aq)$

(e) $2 Au(s) + 3 Cl_2(g) \rightarrow 2 AuCl_3(s)$

20.3 For the earlier part of the Period 4 transition metals, the maximum oxidation number is the same as the Group number (that is, the sum of the $4s$ and $3d$ electrons). For the later members, the oxidation state of +2 predominates.

20.5 Titanium(IV) chloride vaporizes readily, a characteristic of a covalent compound. Covalent behavior would be expected on the basis of the high charge density of the Ti^{4+} ion resulting in a strong attraction for the large, low charge density chloride ions (that is, covalent bond formation).

20.7 (a) $MnO_4^-(aq) + 8 H^+(aq) + 5 e^- \rightarrow Mn^{2+}(aq) + 4 H_2O(l)$

(b) $MnO_4^-(aq) + 2 H_2O(l) + 3 e^- \rightarrow MnO_2(s) + 4 OH^-(aq)$

20.11 The chemical equation is

$$Ni(s) + 4 CO(g) \rightarrow Ni(CO)_4(g)$$

Formation of this compound requires a net decrease of three moles of gas, resulting in a decrease in entropy. Hence the forward reaction must be enthalpy driven (that is, exothermic). The decomposition reaction is favored at higher temperatures. This is logical when the relationship $\Delta G° = \Delta H° - T\Delta S°$ is considered. For the reverse reaction, both $\Delta H°$ and $\Delta S°$ are positive, thus as the temperature increases, $(-T\Delta S°)$ will become increasing negative, to the point where the term exceeds $\Delta H°$ and the reverse reaction becomes spontaneous (negative $\Delta G°$).

20.13 (a) Cobalt:
$$[Co(OH_2)_6]^{2+}(aq) + 4\ Cl^-(aq) \rightarrow [CoCl_4]^{2-}(aq) + 6\ H_2O(l)$$
(b) Copper:
$$2\ Cu(s) + 2\ H^+(aq) + 4\ Cl^-(aq) \rightarrow 2\ [CuCl_2]^-(aq) + H_2(g)$$
$$[CuCl_2]^-(aq) \rightarrow CuCl(s) + Cl^-(aq)$$
(c) Chromium:
$$2\ CrO_4{}^{2-}(aq) + 2\ H^+(aq) \rightarrow Cr_2O_7{}^{2-}(aq) + H_2O(l)$$

20.15 Bromide ion:
$$Ag^+(aq) + Br^-(aq) \rightarrow AgBr(s)$$
$$AgBr(s) + 2\ NH_3(aq) \rightarrow [Ag(NH_3)_2]^+(aq) + Br^-(aq)$$

20.17 A halide ion is an obvious choice because a large negative weak field ligand will favor a tetrahedral arrangement. Iodide would be the specific choice because it is the largest of the halides. Also, vanadium(II) is a very low oxidation state for the metal. To stabilize it, a reducing anion is preferable—and iodide is strongly reducing.

20.19 The amphoteric metals are those which are on the metal boundary with the semimetals and nonmetals, such as zinc and tin. Aluminum fits in this category, while iron(III) is well away from the boundary.

20.21 Oxygen. Iron(III) is a hard acid while oxygen is a hard base (and sulfur a soft base).

20.23 Chromium(VI) oxide should be acidic. It is the metal in the higher oxidation state (that is, with the more oxygens) that will exhibit the more acidic properties.

20.25 Fluoride ion; $[CoF_6]^{3-}$.

20.27 The color must be a result of charge-transfer interactions. Thus for silver iodide it would be
$$Ag^+I^- \leftrightarrow Ag^0I^0$$

20.29 The major similarity was the +1 oxidation state, though for copper, +2 is more common. In terms of chemistry, there are more differences than similarities; for example, the alkali metals are very reactive while the coinage metals are very unreactive.

20.31 (a) Silver:

$$Ag^+(aq) + Cl^-(aq) \rightarrow AgCl(s)$$
$$2\ Ag^+(aq) + CrO_4^{2-}(aq) \rightarrow Ag_2CrO_4(s)$$

(b) Cobalt(II):

$$[Co(OH_2)_6]^{2+}(aq) + 4\ Cl^-(aq) \rightarrow [CoCl_4]^{2-}(aq) + 6\ H_2O(l)$$
$$Co^{2+}(aq) + 2\ OH^-(aq) \rightarrow Co(OH)_2(s)$$

(c) Bromide:

$$Ag^+(aq) + Br^-(aq) \rightarrow AgBr(s)$$
$$2\ Br^-(aq) + Cl_2(aq) \rightarrow Br_2(aq) + 2\ Cl^-(aq)$$

(d) Chromate:

$$CrO_4^{2-}(aq) + Ba^{2+}(aq) \rightarrow BaCrO_4(s)$$
$$2\ CrO_4^{2-}(aq) + 2\ H^+(aq) \rightarrow Cr_2O_7^{2-}(aq) + H_2O(l)$$
$$CrO_4^{2-}(aq) + 8\ H^+(aq) + 3\ e^- \rightarrow Cr^{3+}(aq) + 4\ H_2O(l)$$
$$SO_2(aq) + 2\ H_2O(l) \rightarrow SO_4^{2-}(aq) + 4\ H^+(aq) + 2\ e^-$$

(e) Copper(II):

$$Cu^{2+}(aq) + Zn(s) \rightarrow Cu(s) + Zn^{2+}(aq)$$
$$[Cu(OH_2)_6]^{2+}(aq) + 4\ NH_3(aq) \rightarrow [Cu(NH_3)_4]^{2+}(aq) + 6\ H_2O(l)$$

20.33 Titanium

$$TiO_2(s) + 2\ C(s) + 2\ Cl_2(g) \xrightarrow{\Delta} TiCl_4(g) + 2\ CO(g)$$
$$TiCl_4(g) + O_2(g) \xrightarrow{\Delta} TiO_2(s) + 2\ Cl_2(g)$$
$$TiCl_4(g) + 2\ Mg(l) \xrightarrow{\Delta} Ti(s) + 2\ MgCl_2(l)$$

Vanadium

$$[H_2VO_4]^-(aq) + 4\ H^+(aq) + e^- \rightarrow VO^{2+}(aq) + 3\ H_2O(l)$$
$$VO^{2+}(aq) + 2\ H^+(aq) + e^- \rightarrow V^{3+}(aq) + H_2O(l)$$
$$[V(OH_2)_6]^{3+}(aq) + e^- \rightarrow [V(OH_2)_6]^{2+}(aq)$$

Chromium

$$CrO_4^{2-}(aq) + 2\ Ag^+(aq) \rightarrow Ag_2CrO_4(s)$$

$$CrO_4^{2-}(aq) + H_2O(l) \leftrightarrow HCrO_4^-(aq) + OH^-(aq)$$

$$2\ CrO_4^{2-}(aq) + 2\ H^+(aq) \rightarrow Cr_2O_7^{2-}(aq) + H_2O(l)$$

$$Cr_2O_7^{2-}(aq) + 2\ NH_4^+(aq) \rightarrow (NH_4)_2Cr_2O_7(s)$$

$$(NH_4)_2Cr_2O_7(s) \rightarrow Cr_2O_3(s) + N_2(g) + 4\ H_2O(l)$$

$$Cr_2O_7^{2-}(aq) + 14\ H^+(aq) + 6\ e^- \rightarrow 2\ Cr^{3+}(aq) + 7\ H_2O(l)$$

$$Cr_2O_7^{2-}(aq) + 2\ K^+(aq) \rightarrow K_2Cr_2O_7(s)$$

$$K_2Cr_2O_7(s) + H_2SO_4(aq) \rightarrow 2\ CrO_3(s) + K_2SO_4(aq) + H_2O(l)$$

$$K_2Cr_2O_7(s) + 4\ NaCl(s) + 6\ H_2SO_4(l) \rightarrow 2\ CrO_2Cl_2(l) + 2\ KHSO_4(s)$$
$$+ NaHSO_4(s) + 3\ H_2O(l)$$

$$CrO_2Cl_2(l) + 4\ OH^-(aq) \rightarrow CrO_4^{2-}(aq) + 2\ Cl^-(aq) + 2\ H_2O(l)$$

$$Cr_2O_7^{2-}(aq) + 14\ H^+(aq) + 6\ e^- \rightarrow 2\ Cr^{3+}(aq) + 7\ H_2O(l)$$

$$2\ Cr^{3+}(aq) + Zn(s) \rightarrow 2\ Cr^{2+}(aq) + Zn^{2+}(aq)$$

$$2\ Cr^{2+}(aq) + 4\ CH_3COO^-(aq) + 2\ H_2O(l) \rightarrow Cr_2(CH_3COO)_4(OH_2)_2(s)$$

Manganese

$$MnO_4^-(aq) + e^- \rightarrow MnO_4^{2-}(aq)$$

$$MnO_4^{2-}(aq) + 2\ H_2O(l) + 2\ e^- \rightarrow MnO_2(s) + 4\ OH^-(aq)$$

$$MnO_4^-(aq) + 2\ H_2O(l) + 3\ e^- \rightarrow MnO_2(s) + 4\ OH^-(aq)$$

$$MnO_4^-(aq) + 8\ H^+(aq) + 5\ e^- \rightarrow Mn^{2+}(aq) + 4\ H_2O(l)$$

$$Mn^{2+}(aq) + 2\ OH^-(aq) \rightarrow Mn(OH)_2(s)$$

$$Mn(OH)_2(s) + OH^-(aq) \rightarrow MnO(OH)(s) + H_2O(l) + e^-$$

Iron

$$[Fe(OH_2)_6]^{3+}(aq) + SCN^-(aq) \rightarrow [Fe(SCN)(OH_2)_5]^{2+}(aq) + H_2O(l)$$

$$[Fe(OH_2)_6]^{3+}(aq) + 4\ Cl^-(aq) \leftrightarrow [FeCl_4]^-(aq) + 6\ H_2O(l)$$

$$Fe^{3+}(aq) + 3\ OH^-(aq) \rightarrow FeO(OH)(s) + H_2O(l)$$

$$[Fe(OH_2)_6]^{3+}(aq) + e^- \rightarrow [Fe(OH_2)_6]^{2+}(aq)$$

$$Fe^{3+}(aq) + 2\ S_2O_3^{2-}(aq) \rightarrow [Fe(S_2O_3)_2]^-(aq)$$

$$[Fe(S_2O_3)_2]^-(aq) + Fe^{3+}(aq) \rightarrow 2\ Fe^{2+}(aq) + S_4O_6^{2-}(aq)$$

$$Fe^{2+}(aq) + 2\ OH^-(aq) \rightarrow Fe(OH)_2(s)$$

$$[Fe(OH_2)_6]^{2+}(aq) + NO(aq) \rightarrow [Fe(NO)(OH_2)_5]^{2+}(aq) + H_2O(l)$$

$$Fe(OH)_2(s) + OH^-(aq) \rightarrow FeO(OH)(s) + H_2O(l) + e^-$$

$$Fe^{2+}(aq) + 2\ e^- \rightarrow Fe(s)$$

$2 \text{ Fe}(s) + 3 \text{ Cl}_2(g) \rightarrow 2 \text{ FeCl}_3(s)$

$\text{Fe}(s) + 2 \text{ HCl}(g) \rightarrow \text{FeCl}_2(s) + \text{H}_2(g)$

Cobalt

$[\text{Co(OH}_2)_6]^{3+}(aq) + \text{e}^- \rightarrow [\text{Co(OH}_2)_6]^{2+}(aq)$

$[\text{Co(OH}_2)_6]^{2+}(aq) + 4 \text{ Cl}^-(aq) \rightarrow [\text{CoCl}_4]^{2-}(aq) + 6 \text{ H}_2\text{O}(l)$

$\text{Co}^{2+}(aq) + 2 \text{ OH}^-(aq) \rightarrow \text{Co(OH)}_2(s)$

$\text{Co(OH)}_2(s) + 2 \text{ OH}^-(aq) \rightarrow \text{Co(OH)}_4{}^{2-}(aq)$

$\text{Co(OH)}_2(s) + \text{OH}^-(aq) \rightarrow \text{CoO(OH)}(s) + \text{H}_2\text{O}(l) + \text{e}^-$

$[\text{Co(OH}_2)_6]^{2+}(aq) + 6 \text{ NH}_3(aq) \rightarrow [\text{Co(NH}_3)_6]^{2+}(aq) + 6 \text{ H}_2\text{O}(l)$

$[\text{Co(NH}_3)_6]^{2+}(aq) \rightarrow [\text{Co(NH}_3)_6]^{3+}(aq) + \text{e}^-$

$\text{O}_2(g) + 2 \text{ H}_2\text{O}(l) + 4 \text{ e}^- \rightarrow 4 \text{ OH}^-(aq)$

Nickel

$\text{Ni(CO)}_4(g) \rightarrow \text{Ni}(s) + 4 \text{ CO}(g)$

$\text{Ni}(s) \rightarrow \text{Ni}^{2+}(aq) + 2 \text{ e}^-$

$[\text{Ni(OH}_2)_6]^{2+}(aq) + 4 \text{ Cl}^-(aq) \rightarrow [\text{NiCl}_4]^{2-}(aq) + 6 \text{ H}_2\text{O}(l)$

$\text{Ni}^{2+}(aq) + 2 \text{ OH}^-(aq) \rightarrow \text{Ni(OH)}_2(s)$

$[\text{Ni(OH}_2)_6]^{2+}(aq) + 6 \text{ NH}_3(aq) \rightarrow [\text{Ni(NH}_3)_6]^{2+}(aq) + 6 \text{ H}_2\text{O}(l)$

Copper

$2 \text{ Cu}(s) + 2 \text{ H}^+(aq) + 4 \text{ Cl}^-(aq) \rightarrow 2 \text{ [CuCl}_2]^-(aq) + \text{H}_2(g)$

$\text{Cu}^{2+}(aq) + \text{Zn}(s) \rightarrow \text{Cu}(s) + \text{Zn}^{2+}(aq)$

$\text{Cu}(s) \rightarrow \text{Cu}^{2+}(aq) + 2 \text{ e}^-$

$[\text{Cu(OH}_2)_6]^{2+}(aq) + 4 \text{ NH}_3(aq) \rightarrow [\text{Cu(NH}_3)_4]^{2+}(aq) + 6 \text{ H}_2\text{O}(l)$

$[\text{Cu(OH}_2)_6]^{2+}(aq) + 4 \text{ Cl}^-(aq) \rightarrow [\text{CuCl}_4]^{2-}(aq) + 6 \text{ H}_2\text{O}(l)$

$\text{Cu}^{2+}(aq) + 2 \text{ OH}^-(aq) \rightarrow \text{Cu(OH)}_2(s)$

$\text{Cu(OH)}_2(s) + 2 \text{ OH}^-(aq) \rightarrow [\text{Cu(OH)}_4]^{2-}(aq)$

$\text{Cu(OH)}_2(s) \rightarrow \text{CuO}(s) + \text{H}_2\text{O}(l)$

Beyond the Basics

20.35 $Au^+(aq) + e^- \rightarrow Au(s)$ $E^° = +1.68$ V

   $Au(s) + 2\ CN^-(aq) \rightarrow Au(CN)_2^-(aq)$ $E^° = +0.60$ V

   $Au^+(aq) + 2\ CN^-(aq) \rightarrow Au(CN)_2^-(aq)$ $E^° = +2.28$ V

   $\Delta G^° = -nFE^° = -RTlnK$

$$\ln K = \frac{nFE^°}{RT} = \frac{1 \times (9.65 \times 10^4\ C \cdot mol^{-1}) \times (+2.28\ V)}{8.31\ V \cdot C \cdot mol^{-1} \cdot K^{-1} \times 298\ K} = 88.8$$

   $K = 3.7 \times 10^{38}$

20.37   First we must calculate the two lattice energies:

$$U_{CuF} = -\frac{(6.02 \times 10^{23}\ mol^{-1}) \times 1.638 \times 1 \times 1 \times (1.602 \times 10^{-19}\ C)^2}{4 \times 3.142 \times (8.854 \times 10^{-12}\ C^2 \cdot J^{-1} \cdot m^{-1})(2.08 \times 10^{-10}\ m)}\left(1 - \frac{1}{8}\right)$$

$$= -956\ kJ \cdot mol^{-1}$$

$$U_{CuF_2} = -\frac{(6.02 \times 10^{23}\ mol^{-1}) \times 2.408 \times 2 \times 1 \times (1.602 \times 10^{-19}\ C)^2}{4 \times 3.142 \times (8.854 \times 10^{-12}\ C^2 \cdot J^{-1} \cdot m^{-1})(2.04 \times 10^{-10}\ m)}\left(1 - \frac{1}{8}\right)$$

$$= -2868\ kJ \cdot mol^{-1}$$

Then we set up Born-Haber cycles:

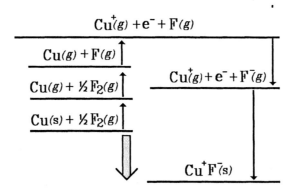

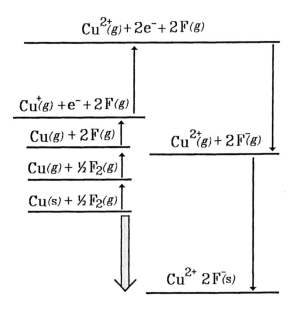

$\Delta H_f^{\circ}(CuF(s)) = [(+337) + \frac{1}{2}(155) + (752) + (-328) + (-956)]$ kJ·mol^{-1}
$= -118$ kJ·mol^{-1}

$\Delta H_f^{\circ}(CuF_2(s)) =$
$[(+337) + (155) + (752) + (+1964) + 2(-328) + (-2868)]$ kJ·mol^{-1}
$= -316$ kJ·mol^{-1}

$Cu(s) + \frac{1}{2} F_2(g) \rightarrow CuF(s)$

$Cu(s) + F_2(g) \rightarrow CuF_2(s)$

In each case, the entropy change will be negative, so the enthalpy decrease needs to be sufficent to overcome the entropy factor.

20.39 $[Co(OS(CH_3)_2)_6]^{2+}\cdot 2ClO_4^-$ and $[Co(OS(CH_3)_2)_6]^{2+}\cdot[CoCl_4]^{2-}$.

20.41   From Appendix,

$VO^{2+}(aq) + 2 H^+(aq) + e^- \rightarrow V^{3+}(aq) + H_2O(l)$ $E^{\circ} = +0.337$ V
  $\Delta G^{\circ} = -1(F)(+0.337 \text{ V}) = -0.337F$

$V^{3+}(aq) + e^- \rightarrow V^{2+}(aq)$ $E^{\circ} = -0.255$ V
  $\Delta G^{\circ} = -1(F)(-0.225 \text{ V}) = +0.225F$

$VO^{2+}(aq) + 2 H^+(aq) + 2 e^- \rightarrow V^{2+}(aq) + H_2O(l)$
  $\Delta G^{\circ} = -0.112F = -2FE^{\circ}$, $E^{\circ} = +0.056$ V

$2 Cr^{2+}(aq) \rightarrow 2 Cr^{3+}(aq) + 2 e^-$ $E^{\circ} = +0.424$ V

$VO^{2+}(aq) + 2 H^+(aq) + 2 e^- \rightarrow V^{2+}(aq) + H_2O(l)$ $E^{\circ} = +0.056$ V

$VO^{2+}(aq) + 2 Cr^{2+}(aq) + 2 H^+(aq) \rightarrow V^{2+}(aq) + 2 Cr^{3+}(aq) + H_2O(l)$ $E^{\circ} = +0.480$ V

$\Delta G^\circ = -n FE^\circ = -RTlnK$

$$\ln K = \frac{n FE^\circ}{RT} = \frac{1 \times (9.65 \times 10^4 \text{ C} \cdot \text{mol}^{-1}) \times (+0.480 \text{ V})}{8.31 \text{ V} \cdot \text{C} \cdot \text{mol}^{-1} \cdot \text{K}^{-1} \times 298 \text{ K}} = 18.7$$

$K = 1.3 \times 10^8$

20.43 Fluorine tends to promote metals to their highest oxidation states, because it is a strong oxidizing agent, whereas iodide ion is a reducing agent; thus the most likely product with fluorine is WF_6.

20.45 Like the equivalent iron compound, this is probably nickel(II) disulfide(2−)—the disulfide ion being analogous to the dioxide(2−) ion. The sulfur oxidation state would be −1.

20.47 [A] Nickel(II) sulfide, [B] hydrogen sulfide, [C] hexaaquanickel(II) ion, [D] sulfur dioxide, [E] sulfur, [F] and [G] disulfur dichloride and sulfur dichloride, [H] hexaamminenickel(II) ion, [I] nickel(II) hydroxide, [J] nickel metal,
[K] tetracarbonylnickel(0).

$NiS(s) + 2 H^+(aq) \rightarrow Ni^{2+}(aq) + H_2S(g)$

$2 H_2S(g) + 3 O_2(g) \rightarrow 2 H_2O(l) + 2 SO_2(g)$

$2 H_2S(g) + SO_2(g) \rightarrow 2 H_2O(l) + 3 S(s)$

$2 S(s) + Cl_2(g) \rightarrow S_2Cl_2(l)$

$S(s) + Cl_2(g) \rightarrow SCl_2(l)$

$[Ni(OH_2)_6]^{2+}(aq) + 6 NH_3(aq) \rightarrow [Ni(NH_3)_6]^{2+}(aq) + 6 H_2O(l)$

$Ni^{2+}(aq) + 2 OH^-(aq) \rightarrow Ni(OH)_2(s)$

$Ni^{2+}(aq) + Zn(s) \rightarrow Ni(s) + Zn^{2+}(aq)$

$Ni(s) + 4 CO(g) \rightarrow Ni(CO)_4(l)$

20.49 Vanadium is the only transition metal whose chemistry matches the information provided.

20.51 Chromium-52 has 28 neutrons corresponding to a full neutron shell.

20.53 Silver chloride, bromide, and iodide are insoluble, silver fluoride is soluble. Calcium chloride, bromide, and iodide are soluble, while calcium fluoride is insoluble. Silver is a large, low charge (low charge density) cation; thus it will form more stable lattices with large, low charge anions, like the chloride to iodide members of the series. These compounds will therefore be insoluble. Calcium, on the other hand, has a higher charge density. It will form a more stable lattice with the smallest of the halogens, fluoride.

20.55 Assume 1.00 cm^3 ($= 1.00$ mL) of metal. Density of metal $= 12.0$ g·cm^{-3}
Mass of metal $= (1.00$ cm$^3)(12.0$ g·cm$^{-3}) = 12.0$ g
Mole of palladium $= (12.0$ g$)/(106.4$ g·mol$^{-1}) = 0.113$ mol Pd
Volume of hydrogen $= 935$ mL
Mole H$_2$ $=$ PV/RT $= (100$ kPa$)(0.935$ L$)/(8.31$ kPa·L·mol^{-1}·K$^{-1})(298$ K$)$
$\qquad = 0.0378$ mol
Ratio 0.113 mol Pd: 0.0378 mol H$_2$ $= 3$ Pd: 2H$_2$
Formula $=$ Pd$_3$H$_2$
Mass H$_2$ $= (0.0378$ mol$)(1.01$ g·mol$^{-1}) = 0.0762$ g H$_2$
Density $= (0.0762$ g$)/(1.00$ cm$^3) = 0.0762$ g·cm^{-3}
Thus the density of dihydrogen within the palladium is almost identical to that in liquid dihydrogen.

20.57 Under normal acid conditions, manganese is reduced from +7 to +2 oxidation states:
MnO$_4^-$(aq) + 8 H$^+$(aq) + 5 e$^-$ → Mn^{2+}(aq) + 4 H$_2$O(l)
Fluoride stabilizes high oxidation states, thus it would not be unreasonable to propose that reduction occurs only down to the +3 oxidation state:
MnO$_4^-$(aq) + 6 F$^-$(aq) + 8 H$^+$(aq) + 4 e$^-$ → [MnF$_6$]$^{3-}$(aq) + 4 H$_2$O(l)
This would require $(20$ mL$) \times (^5/_4) = 25$ mL of titrant to oxidize the iron(II) ion.

20.59 CuCN(s) + CN$^-$(aq) → [Cu(CN)$_2$]$^-$(aq)

20.61 (a) MnO: [Mn] + [–2] = 0, Mn = +2
Mn$_3$O$_4$: 3[Mn] + 4[–2] = 0, Mn = $+^8/_3$
Mn$_2$O$_3$: 2[Mn] + 3[–2] = 0, Mn = +3
MnO$_2$: [Mn] + 2[–2] = 0, Mn = +4
Mn$_2$O$_7$: 2[Mn] + 7[–2] = 0, Mn = +7

(b) Mn_3O_4 is a mixed oxide containing $(Mn^{2+})(Mn^{3+})_2(O^{2-})_4$.

(c) MnO should be basic and Mn_2O_7 acidic. Low oxidation state metal oxides are basic and high oxidation state, acidic.

(d) MnO_2 is the analog of ClO_2 (Mn_2O_7 is the analog of Cl_2O_7 but neither of these are common oxides).

20.63 (a) $Fe(s) + O_2(g) \rightarrow Fe_2O_3(s)$

(b) The layer of unreactive sodium silicate will coat the iron and prevent the continuation of the oxidation.

(c) The red-hot iron would have reacted with water to give hydrogen gas.

$$2\ Fe(s) + 3\ H_2O(l) \xrightarrow{\Delta} Fe_2O_3(s) + 3\ H_2(g)$$

The explosion would have resulted from the hydrogen/oxygen mixture:

$$2\ H_2(g) + O_2(g) \rightarrow 2\ H_2O(g)$$

Chapter 21

THE GROUP 12 ELEMENTS

Exercises

21.1 (a) $Zn(s) + Br_2(l) \rightarrow ZnBr_2(s)$

(b) $ZnCO_3(s) \rightarrow ZnO(s) + CO_2(g)$

21.3 React zinc metal with a dilute acid, such as hydrochloric acid, then add a soluble carbonate, such as sodium carbonate:

$$Zn(s) + 2\,H^+(aq) \rightarrow Zn^{2+}(aq) + H_2(g)$$
$$Zn^{2+}(aq) + CO_3^{2-}(aq) \rightarrow ZnCO_3(s)$$

21.5 (a) Zinc and magnesium have the following similarities: their cations are 2+ ions of similar size, they are colorless, and they both form hexahydrates. Both elements form soluble chlorides and sulfates, and insoluble carbonates. The anhydrous chlorides are covalently bonded and hydroscopic. However, zinc hydroxide is amphoteric, whereas magnesium hydroxide is basic.

(b) Although both metals form colorless cations, that of zinc has a +2 oxidation state and aluminum has a +3 oxidation state. In fact, the only two common features are that both zinc and aluminum are amphoteric metals, reacting with both acids and bases, and they are both strong Lewis acids, their soluble salts forming strongly acidic solutions.

21.7 The charging reactions will be

$$Cd(OH)_2(s) + 2\,e^- \rightarrow Cd(s) + 2\,OH^-(aq)$$
$$2\,Ni(OH)_2(s) + 2\,OH^-(aq) \rightarrow 2\,NiO(OH)(s) + 2\,H_2O(l) + 2\,e^-$$

21.9 Silver chloride, m.pt. 455°C, thallium(I) chloride, m.pt. 430°C

Silver bromide, m.pt. 430°C, thallium(I) bromide, m.pt. 480°C

21.11 $Zn(s) + 2 H^+(aq) \rightarrow Zn^{2+}(aq) + H_2(g)$

$Zn(OH_2)_6^{2+}(aq) + 4 NH_3(aq) \rightarrow Zn(NH_3)_4^{2+}(aq) + 6 H_2O(l)$

$Zn^{2+}(aq) + 2 OH^-(aq) \rightarrow Zn(OH)_2(s)$

$Zn(OH)_2(s) + 2 OH^-(aq) \rightarrow Zn(OH)_4^{2-}(aq)$

$Zn(OH)_2(s) \rightarrow ZnO(s) + H_2O(l)$

$ZnO(s) + 2 H^+(aq) \rightarrow Zn^{2+}(aq) + H_2O(l)$

$ZnCO_3(s) \rightarrow ZnO(s) + CO_2(g)$

Beyond the Basics

21.13 Mercury(I) undergoes a disproportionation equilibrium:

$Hg_2^{2+}(aq) \rightarrow Hg(l) + Hg^{2+}(aq)$

This equilibrium is driven to the right by the formation of highly insoluble (high lattice energy) compounds such as mercury(II) selenide:

$Hg^{2+}(aq) + Se^{2-}(aq) \rightarrow HgSe(s)$

21.15 For metals to mix, they must have similar size of atoms and adopt the same crystal packing arrangement (see Chapter 4). The metals are very different in size, and it can be found from data tables that they have different packing arrangements (lead is cubic close-packed while zinc is hexagonal close-packed).

21.17 Sulfur. Mercury(II) is a soft acid. Sulfur is a soft base whereas oxygen is a hard base.

21.19 (a) $Zn(NH_2)_2(NH_3) + 2 NH_4^+(NH_3) \rightarrow Zn(NH_3)_4^{2+}(NH_3)$

(b) $Zn(NH_2)_2(NH_3) + 2 NH_2^-(NH_3) \rightarrow Zn(NH_2)_4^{2-}(NH_3)$

21.21 Zinc oxide. The (+2)(−2) ionic attractions will be greater than those between (+2) and (−1) ions. In other words, the lattice energy of zinc oxide will be much greater than that of zinc chloride.

21.23 Hydrogen sulfide is in a two-step equilibrium with the sulfide ion.

$H_2S(aq) + H_2O(l) \leftrightarrow H_3O^+(aq) + HS^-(aq)$

$HS^-(aq) + H_2O(l) \leftrightarrow H_3O^+(aq) + S^{2-}(aq)$

In neutral conditions, $[S^{2-}]$ must be greated than the value necessary for the K_{sp} for zinc sulfide to be exceeded. However, when acidified, the increased hydronium ion concentration will "drive" the equilibria to the left; presumably sufficiently to decrease $[S^{2-}]$ below the necessary value for zinc sulfide precipitation.

Chapter 22

ORGANOMETALLIC CHEMISTRY

22.1   (a)   organometallic

      (b)   not organometallic as the bond B-O not B-C

      (c)   organometallic

      (d)   not organometallic as nitrogen is not metallic

      (e)   not organometallic as there is no Na-C bond

      (f)   organometallic

      (g)   organometallic

22.3   (a) $Bi(CH_3)_5$

      (b) $Si(C_6H_5)_4$

      (c) $KB(C_6H_5)_4$

      (d) $Li_4(CH_3)_4$

      (e) $(C_2H_5)MgCl$

22.5 C_2H_5MgBr will be tetrahedral with two molecules of solvent coordinated to the magnesium. In contrast, the bulky organo group in $2,4,6\text{-}(CH3)_3C_6H_2MgBr$ leads to a coordination number of two.

22.7   A transmetallation reaction involves the breaking of the metal-carbon bond and the forming of a metal-carbon bond to a different metal. Mercury compounds are commonly used in transmetallation to form organometallic compounds of other metals.

$$Hg(CH_3)_2 + 2\,Na \rightarrow 2\,NaCH_3 + Hg$$

22.9   (a) $CH_3Br + 2\,Li \rightarrow LiCH_3 + LiBr$

      (b) $MgCl_2 + 2\,LiC_2H_5 \rightarrow 2\,LiCl + Mg(C_2H_5)_2$

      (c) $Mg + (C_2H_5)_2Hg \rightarrow Mg(C_2H_5)_2 + Hg$

      (d) $C_2H_5Li + C_6H_6 \rightarrow Li(C_6H_5) + C_2H_6$

      (e) $Mg + C_2H_5HgCl \rightarrow C_2H_5MgCl + Hg$

      (f) $B_2H_6 + CH_3CH{=}CH_2 \rightarrow B(CH_2CH_2CH_3)_3$

      (g) $SnCl_4 + 4\,C_2H_5MgCl \rightarrow Sn(C_2H_5)_4 + 4\,MgCl_2$

22.11 (a) hexacarbonylchromium(0)

(b) ferrocene or bis(pentahaptocyclopentadienyl)iron(II)

(c) hexahaptobenzenetricarbonylchrmium(0)

(d) pentahaptocyclopentadienyltricarbonyltungsten(I)

(e) bromopentacarbonylmanganese(I)

22.13 $Cr(CO)_6$

Cr(0) electrons ($3d^6$) $= 6$
CO electrons $= 6 \times 2$ $= 12$
Total $= 18$

$Fe(CO)_5$

Fe(0) electrons ($3d^8$) $= 8$
CO electrons $= 5 \times 2$ $= 10$
Total $= 18$

$Ni(CO)_4$

Ni(0) electrons ($3d^{10}$) $= 10$
CO electrons $= 4 \times 2$ $= 8$
Total $= 18$

22.15 $V(CO)_6$ is a seventeen electron complex. Therefore, it readily gains another electron to give eighteen.

22.17 (a) Mn(0) electrons ($3d^7$) $= 7$
CO electrons $= 5 \times 2$ $= 10$
Total $= 17$ therefore 1 Mn-Mn bond

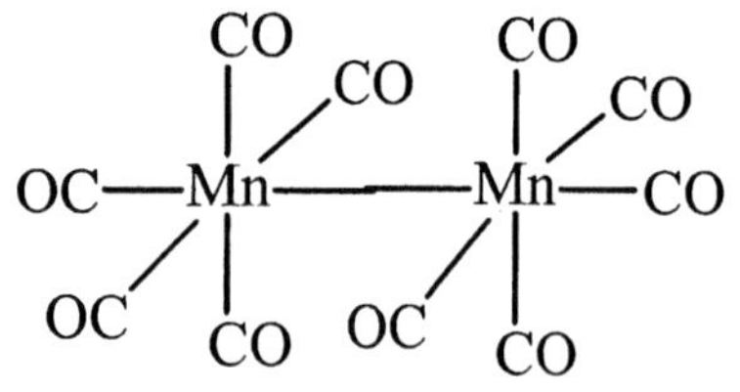

(b)   Mn(I) electrons ($3d^6$)   = 6
        CO electrons = 2 × 2   = 4
        η^5-C_5H_5 electrons   = 6
                            total   = 16   therefore 2 Mn-Mn bonds

(c)   Fe(0) electrons $3d^8$ = 8
        CO electrons = 2 × 2   = 4
        η^4-C_4H_4 electrons   = 4
        μ -CO electrons = 2 ÷2   = 1
                            total   = 17   therefore 1 Fe-Fe bond

(d)   Mn(I) electrons ($3d^6$)   = 6
        CO electrons = 4 × 2   = 8
        μ-Br electrons = 2 × 2   = 4
                            total   = 18   therefore no Mn-Mn bonds

22.19 (a) $Cr(CO)_6 + 3\ CH_3CN \rightarrow Cr(CO)_3(CH_3CN)_3 + 3\ CO$

(b) $Mn_2(CO)_{10} + H_2 \rightarrow 2\ HMn(CO)_5$

(c) $Mo(CO)_6 + (CH_3)_2PCH_2CH_2P(Ph)CH_2CH_2P(CH_3)_2 \rightarrow$
$Mo(CO)_3((CH_3)_2PCH_2CH_2P(Ph)CH_2CH_2P(CH_3)_2) + 3\ CO$

(d) $Fe(CO)_5 + 1,3\text{-cyclohexadiene} \rightarrow 2\ CO + $ $(CO)_3Fe—$

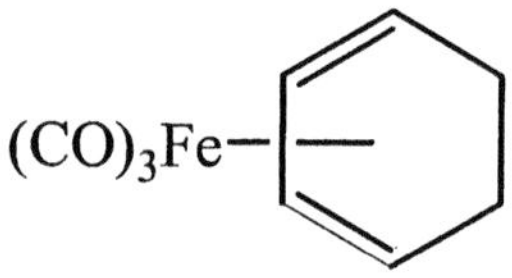

(e)

$NaMn(CO)_5 + CH_2{=}CHCH_2Cl \longrightarrow NaCl + (CO)_5MnCH_2CH{=}CH_2$

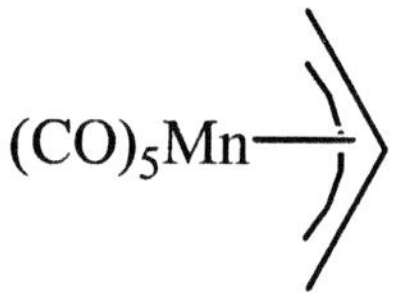

$\longrightarrow (CO)_5Mn— \quad + CO$

(f)

$Cr(CO)_6 + C_6H_6 \longrightarrow (CO)_3Cr— \quad + 3\ CO$

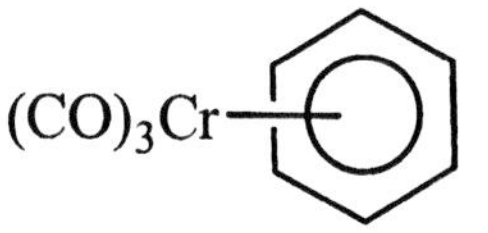

(g)

$PtCl_2(PMe_3)_2 + LiCH_2CH_2CH_2CH_2Li \longrightarrow 2\ LiCl + (PMe_3)_2Pt$

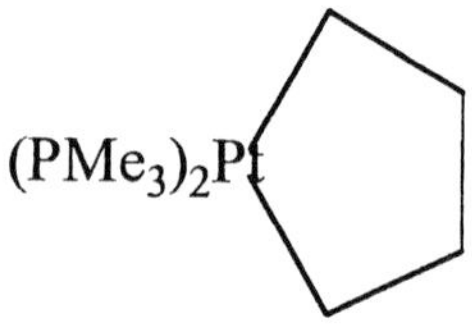

(h) $Ni(CO)_4 + PF_3 \rightarrow Ni(CO)_3PF_3 + CO$

(i) $Mn_2(CO)_{10} + Br_2 \rightarrow 2\ Mn(CO)_5Br$

(j) $HMn(CO)_5 + CO_2 \rightarrow (CO)_5MnCOOH$

22.21 (a)   Iridium (III)
(b)   Charge on H_3^+ would be +1.
H_3^+ would have 2 electrons over the H-H-H bonds. These would be 2 electron 3 center bonds, as in diborane.

Beyond the Basics

22.23

$$(\eta^5\text{-}C_5H_5)_2Ni + Ni(CO)_4 \longrightarrow 2$$

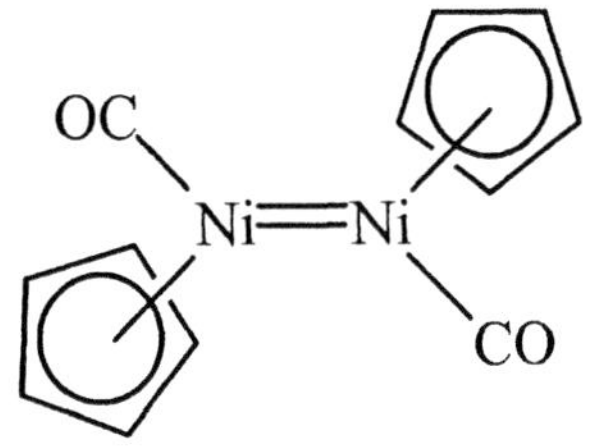

22.25

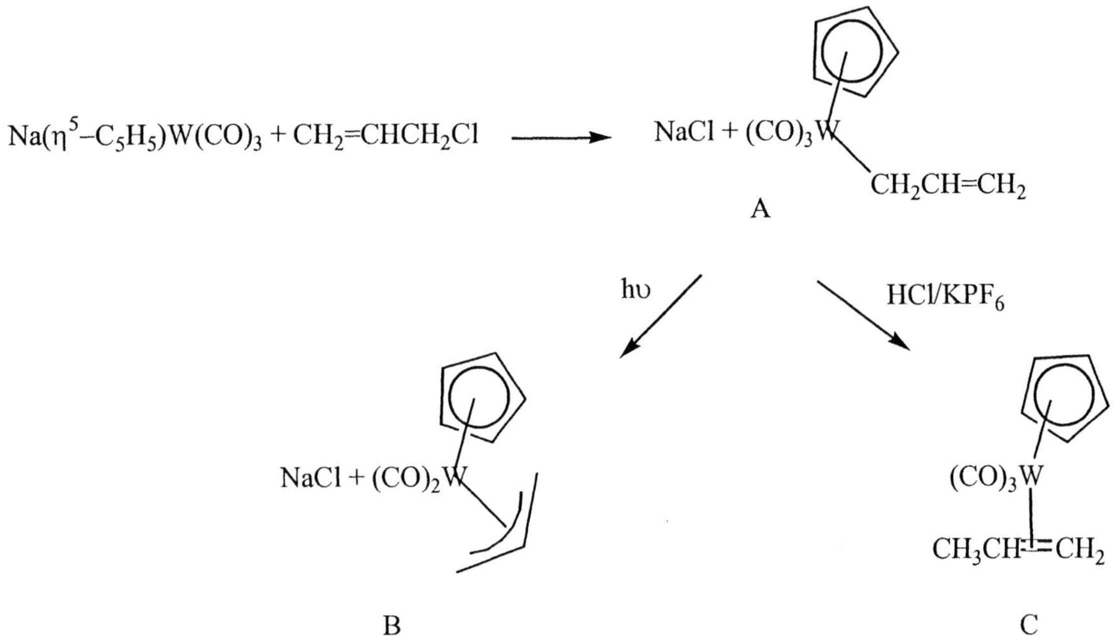

$$Na(\eta^5\text{-}C_5H_5)W(CO)_3 + CH_2=CHCH_2Cl \longrightarrow NaCl + (CO)_3W$$

$CH_2CH=CH_2$

A

hυ / HCl/KPF$_6$

$NaCl + (CO)_2W$

B

$(CO)_3W$

$CH_3CH=CH_2$

C

A = tricarbonyl(η^5-cyclopentadienyl)(η^1-propenyl)tungsten(II)
B = dicarbonyl(η^5-cyclopentadienyl)(η^3-propenyl)tungsten(II)
C = tricarbonyl(η^5-cyclopentadienyl)(η^2-propenyl)tungsten(II)
hexafluorophosphate
Evolved gas = propene

22.27 The reaction of alkylated titanium(IV) with carbon disulphide will give
$Ti(S_2CEt_2)_4$.

Chapter 23

THE RARE EARTH AND ACTINOID ELEMENTS

Exercises

23.1   (a) $2\,Eu(s) + 6\,H_2O(l) \rightarrow 2\,Eu(OH)_3(s) + 3\,H_2(g)$

       (b) $UO_3(s) + H_2SO_4(aq) \rightarrow (UO_2)SO_4(aq) + H_2O(l)$

23.3   If the europium 2+ ion has similar solubilities of its common salts to the strontium ion, we would expect the nitrate and chloride to be soluble and the sulfate and carbonate to be insoluble.

23.5 $[Ce(OH_2)_6]^{4+}(aq) + H_2O(l) \rightarrow [Ce(OH_2)_5(OH)]^{3+}(aq) + H_3O^{+}(aq)$

23.7   The shorter half-lives of actinium and protactinium can be accounted for by the fact that they have odd numbers of protons (see Chapter 2).

23.9   The oxidation states of the early actinoids match with those of the transition metals above them. For example, thorium has a common oxidation number of +4 like titanium, zirconium, and hafnium (compare Figure 21.5 with Table 19.2). Uranium has the interesting parallel with chromium in that it forms the yellow $U_2O_7{}^{2-}$ ion like the orange $Cr_2O_7{}^{2-}$ ion.

Beyond the Basics

23.11   The U^{4+} ion is the most thermodynamically stable while uranium metal is a very strong reducing agent. The $UO_2{}^{2+}$ ion in which uranium has an oxidation state of +6 is not significantly oxidizing as might be expected from the high positive state. The $UO_2{}^{+}$ ion is prone to disproportionation. The U^{3+} is reasonably stable but it can be oxidized to the U^{4+} ion.

23.13 $Mt = [Rn]7s^{2}5f^{14}6d^{7}$.

23.15   They both adopt the same crystal packing arrangement and they have similar metallic radii (Ti = 145 pm, U = 139 pm).